HUMAN ANATOMY AND PHYSIOLOGY

POSTURE

TYPES, EXERCISES AND HEALTH EFFECTS

HUMAN ANATOMY AND PHYSIOLOGY

Additional books in this series can be found on Nova's website
under the Series tab.

Additional e-books in this series can be found on Nova's website
under the e-book tab.

HUMAN ANATOMY AND PHYSIOLOGY

POSTURE

TYPES, EXERCISES AND HEALTH EFFECTS

SARAH A. CURRAN
EDITOR

New York

For permission to use material from this book please contact us:
Telephone 631-231-7269; Fax 631-231-8175
Web Site: http://www.novapublishers.com

Library of Congress Cataloging-in-Publication Data

ISBN: 978-1-63117-252-6

Library of Congress Control Number: 2014931436

Published by Nova Science Publishers, Inc. † New York

Contents

Preface

"Movement is a medicine for creating change in a person's physical, emotional, and mental states".

Carol Welch

One of our most precious possessions for each of us is our body. It performs a series of automatic, complex and important mechanisms such as breathing, digestion, and ambulating from one point to the next with little conscious effort. Yet, as a consequence of new experiences or when circumstances change, such as when an individual becomes ill or injured the body comes under volitional control. Whilst unique to humans, bipedal stance creates an inherently unstable platform which can challenge balance and the neuromuscular control system. Integrated within this is posture, which is described as the position of the human body and its orientation in space. A more specific definition is provided by Britnell et al. (1) who suggest that posture is related to musculoskeletal alignment and balance that acts to protect the human body from injury, progressive deformity and create efficiency. Nevertheless, movement and posture of the human body changes with the routines of daily life and can be influenced by a variety of factors that include fatigue, general health, state of mind, participation in physical activity, and musculoskeletal imbalance and malignment.

In the last 25 years, the quest for knowledge and understanding on the role of posture has been constant and continues to evolve. In particular, in recent years there has been a growing recognition of psychosomatic conditions on the influence posture. This however adds further to the complexity of understanding the variability of physical impairments and compensation mechanisms each individual may present with. In essence, the clinician must be reflective and perceptive, and acknowledge that there is no 'one fix' treatment strategy of managing postural related conditions and pathologies.

This book presents a collection of work on the various aspects of posture, including its evolutionary trend of knowledge and its rehabilitation. Each invited international author has contributed significantly in their field and are passionate for continued learning and understanding that integrates theory into clinical practice. The book provides a timely reminder to recognise the importance of a healthy and active lifestyle that is synonymous with modern day living, yet in turn acknowledge the consequences of activity (potential for injury) and value of exercise on improving posture. The book also provides an opportunity to appraise the role of yoga and Pilates within rehabilitation programs. This inclusion (along with other and traditional forms of rehabilitation protocols) echoes the increasing trend of individuals participating in yoga and Pilates across the world. To reiterate this point, in 2012, over 20 million Americans practiced yoga, which represents an increase of 29% from 2008. In contrast, Pilates, which was introduced by Joseph Pilates in the late 1960s and was often mistaken for the training of pilots has received an incremental stream of followers. However, it was not until the arrival of the noughties till it gained a rapid growth in popularity.

The book presents twelve chapters and opens with a chapter by Rosário which provides a contemporary review based on various optical systems for postural assessment. In the first of his four chapter contributions, Rosário explores the role of three dimensional and two dimensional analysis and note the issues related to cost, time, space availability, as well as marker placement and other sources of errors. He also discusses the role of smart phones and contribution of three dimensional sensors – made popular by the introduction of Microsoft® *Kinect* in association with the X-box 360 games console. Rosário also discusses the introduction of a new and more accurate type of sensor referred to as 'time of flight' for the new generation of *Kinect* which may appeal to postural assessment within a clinical setting. The clinical usefulness of head posture assessment for patients presenting with neck pain is presented by Silva et al. in chapter two. Using an accumulation of recent research and their own findings, the authors set out to determine operational definitions for ideal, normal and abnormal head posture. The methods used to assess head posture are then explored and the relationships between head posture and neck pain are comprehensively reviewed. In particular, the authors note the value of forward head posture assessment for patients with neck pain, but not for head extension/flexion, side-flexion and rotation. This observation is based on the lack of evidence used to explore these postural components and the authors acknowledge the need for further research to enhance the understanding, assessment and management regimens of head posture and neck pain. The third chapter is written by Sonnesen who explores the relationship between head and neck posture and upper spine morphology to the craniofacial profile and orofacial function. The chapter which provides an appreciation of the standard procedure used to record head and neck posture and upper spine morphology provides an evidenced-based approach of previous literature. The author points out that head and neck posture as well as upper spine morphology can influence the developmental of the craniofacial profile and orofacial function for temporomandibular disorders as well as obstruction to the upper air ways. These factors are of clinical value for a range of health professionals which include the dentist, orthodontist, general physician and physiotherapist.

In an appreciation of the vast array of biomechanical, environmental, neurophysiology, pathological, physiological and psychological conditions that can influence posture, Rosário presents two chapters, the first of which is based on acknowledging the role of emotion and its link with posture. He states the need to approach the management of postural related conditions from a multi-dimensional purpose, which should include the role of emotions. In particular, Rosário points out the link between pain and its influence on emotion and posture. He also shows how the role of body image and the effects of emotional tension including depression, anger and sadness can have on posture. Rosário's third chapter, and chapter five of the book, builds on from the previous chapter and reports on the findings of a study which investigated the relationship between body posture on subjective fear, concern and happiness in adult women. Using a digital camera, each volunteer was photographed standing from a lateral (right angle) and anterior view. A series of angles were obtained to determine postural assessment (alignment) from each view. Additional data was obtained from a visual analogue scale to document the degree of subjective depression. Rosário shows how current fear, usual concern, and current and usual happiness is related to various postural alterations. Whilst it is acknowledged that further research is required, Rosario notes that the findings of this data may be useful clinically in the identification of a patient's emotional status from a non-verbal perspective.

The next three chapters are based on a range of sporting activities, the first of which is presented by Minarro et al. who explore the influence of hamstring extensibility on spinal and pelvic posture in highly trained athletes. From a sample of 260 paddlers, three groups were created based on the passive straight leg angle raise (group 1 <73°; group 2 ≥73 and <84°; group 3 ≥84°). The Spinal Mouse was used to record sagittal plane spinal curvatures and pelvic inclination during various test conditions that included standing, maximal trunk flexion and knee extended (sit-and-reach test) and flexed. A greater thoracic and posterior pelvic tilt during maximum trunk flexion was noted with the knees extended and flexed. These observations have the potential to increase the load on the spine and as a consequence support the need to integrate a systematic stretching program for the hamstrings. Chapter seven is a comprehensive review written by Muyor et al. and is devoted to the popular sport of cycling and how it can influence spinal posture and predispose to injury and pain. The authors discuss the relationship between cycling and low back pain, and spinal posture whilst on the bicycle. They also note the comparisons of spine morphology of the cyclist during standing and in various trunk flexed positions. Based on the literature reviewed, the authors also discuss the influence of age on angular alignment of cyclists. The authors highlight the role of core stability and strengthening in cycling for improving posture whilst cycling and preventing low back pain in cyclists. The effects of physical and sporting activities on postural activity in children are explored by Sahli et al. in chapter eight. Here the authors highlight the anxieties related to the potential detrimental effect on postural stability and control, and its relationship to physical and sporting activities, and injury. The authors provide a comprehensive review and set the scene by exploring the effects of physical and sporting training effects on postural control in adults. This is supported by a discussion on postural development control in children that is followed by exploring the effects of physical and sporting training on postural stability in children. From the literature explored, they note that female children who took part in gymnastics and circus related activities improved postural stability. Whilst the authors note further research is required, acknowledgment of the relationship between physical and sporting activities on postural stability for clinicians may be useful for incorporating rehabilitation programmes.

Chapter nine considers the role of unstable shoes for the improvement postural control. Presented by Sousa and Tavares, this informative review considers how the neuromuscular system alters whilst standing in unstable footwear. What is particularly important in this chapter is the appreciation of the short and long term postural control reorganisation effects to the wearing of unstable shoes, and how they have the potential to be integrated into a rehabilitation protocol to improve postural performance. The last three chapters are related to use of yoga and Pilates for the management of various postures and control of postural stability. For his final contribution in the tenth chapter, Rosário presents the efficacy of postural chains therapy that employs modified yoga positions for spinal treatment. The chapter presents a study of 200 individuals with spinal related complaints. Two groups (n = 100/group) were formed, with one receiving two 15 minute treatments based on the muscular postural chain therapy (modified yoga) and the other group (control) receiving two 15 minutes relaxation (supine position). It was noted that the group who received the postural chain therapy (modified yoga) produced an immediate reduction in musculoskeletal spinal pain, which was potentially as a result of postural improvement. In a slightly alternative approach, Arya shows how the role of yoga can be used for colon cleanse. This penultimate chapter (eleven) documents how the drinking of saline water and a series of postural changes (5 yoga poses), along with relaxation, deep breathing and meditation can assist with bowel cleansing via parasympathetic activation. This more natural approach negates the need for large amounts of normal saline/electrolyte solutions over a prolonged period of time (4 – 6 hours) and has the potential to appeal to those individuals requiring colon cleanse.

Postural angles in adolescents who wear high heels and the integration of a Pilates as a means of postural control is the focus of the last chapter (chapter twelve) by Pezzan and Lopes. The authors initially explore the role of posture and the influence of growth and development. The influence of high heels on posture in general is also discussed. This is followed by the presentation of findings of a study based on 100 female adolescents. Two groups were created, one based on non-users of high heeled shoes and the other users of high heeled shoes. The first part of the study focused on the use habits of high heeled shoes and noted that the majority of adolescents wore high heeled shoes at some point on a weekly basis. The second part involved analysis of lumbar spine posture, pelvis and lower limbs and a series of alterations were observed with the use of high heeled shoes. The authors note the importance of Pilates and how it could assist in improving postural alignment.

This book is by no means a complete guide to posture and its related factors; however it does capture the evolutionary trend of knowledge on this fascinating subject. Perhaps more importantly, it offers an insight into how methods of rehabilitation are being integrated within clinical practice. Whilst the constant drive towards evidenced-based-practice remains at the core of clinical practice, the need to demonstrate clinical effectiveness remains a priority. It has been a privilege to have worked with a range of international authors with various backgrounds, and each are to be commended and thanked for their valuable contribution. As a final note: the reader should be open-minded and appreciate alternative approaches and management strategies to postural related conditions. So…sit up, check your posture and happy reading!

Dr. Sarah A. Curran
Senior Lecturer
Wales Centre for Podiatric Studies, Cardiff School of Health Sciences
Cardiff Metropolitan University, Western Avenue, Cardiff, CF5 2YB United Kingdom
Tel: +44 (0) 29 2041 7221
scurran@cardiffmet.ac.uk
December, 2013

References

Britnell SJ, Cole JV, Isherwood L, Sran MM, Britnell N, Burgi, S, et al. Postural health in women: the role of physiotherapy. *J Obstet Gynaecol Can*. 2005;27;493-510.

In: Posture: Types, Exercises and Health Effects
Editor: Sarah A. Curran
ISBN: 978-1-63117-252-6

Chapter 1

Old Problems and New Perspectives for Postural Analysis

José Luís Pimentel do Rosário,* PT, PhD
Federal University of São Paulo, São Paulo, Brazil

Abstract

The study of posture is not an easy task, mainly because postural assessment is still scientifically inaccurate. Photographs of bipedalism in the frontal and sagittal planes are one of the most widely used methods for this assessment. Motion capture allows 3D assessment with increased accuracy, but is considered expensive yet with the introduction of new technologies it is becoming more cost effective. This chapter presents a review of the literature which aims to discuss the advantages and disadvantages of the various types of optical systems for postural assessment. Medline and Lilacs databases were searched for the period from 2002 to 2012, using the following key terms: "posture" and "postural". Articles needed to have a description of an optical postural assessment. From the literature explored, it was noted that the advantages of photos are relate to their cost effectiveness and portability. It is even possible to find smart phones apps for postural analysis. Nevertheless, the chances of errors increase with this method. For example, issues in pinpointing body landmarks and finding the correct position of the camera (small rotation can distort the picture) are reported as the two major flaws. They can however be minimized with a better landmark choice and fixed camera positioning. In comparison, motion capture systems have several advantages over pictures. It can assess the patient from different angles simultaneously, assess not just the posture, but also gait and the dynamical part of the posture, which better satisfies the non-stillness postural concept. However, besides the same landmarks problems found in pictures, motion capture systems can be far more expensive and take much more space, and it can be a challenge to transport the system from one venue to the next. A markerless system using a good algorithm can be an option to be tested for accuracy. Whilst 3D sensors such as Microsoft *Kinect* at present can be considered as not the optimal solution, they may be in a near future. They are inexpensive, portable and do not require markers to determine

* Contact email: joerospt@gmail.com

anatomical landmarks, and consequently may overcome the limitations associated with laboratory-based movement analysis systems. The limitation however is the lack of precision, but new technologies such as the time of flight cameras are improving in accuracy. Researchers and clinicians should carefully choose their assessment equipment based on accuracy, pre-assessment preparation time, space available for equipment installation and cost.

Keywords: Assessment, photogrammetry, posture, 2D, 3D, time of flight, markerless system, *Kinect*

Introduction

The study of human posture is relatively new compared to other areas of medical science. Certain deviations in posture can adversely affect muscular efficiency, joint positions and ligaments workload. It can be unsightly as well as predisposing individuals to musculoskeletal pathologic conditions and pain. [1-3] However, it is not an easy subject to study, mainly because postural assessments are still scientifically inaccurate. Two methods are widely used for this assessment: the study of the projection of the center of gravity with the aid of a force platform; [4] and photography of the standing posture, using both frontal and sagittal planes. [3] Other methods, such as magnetic resonance imaging are expensive, while others such as radiographs and computed tomography involve radiation problems. [5-7]

The problem with the first approach is purely semantic. Some studies refer to postural analysis as measured by the force platform, [8] but this is inaccurate. The force platform measures the oscillation of the body and the association between the projection of the center of gravity and the support base, [9] thus providing a balance, not a posture, measurement. There is some scientific evidence that exists to establish connections between posture and equilibrium problems with orthopedic and rheumatologic diseases. These include knee osteoarthritis, ankle instability, neck tension and back pain. [4] Therefore, posture seems to be strongly related to balance [10,11] and its treatment can be similar, but posture is not the same thing as balance. It is very difficult to imagine a person with good posture and poor balance, but it is possible to imagine bad posture with good balance if the misaligned body segments are compensated so that the resulting projection of the center of gravity is between the feet.

The problem with the second approach is that the adhesive markers are not accurate. These are used in the demarcation of specific bone landmarks adopted as the reference point for calculating distances and angles on the photos. Depending on the chosen anatomical region, it is easy to misplace the exact location. Large measurements, such as the distance between the shoulders for example, may not suffer so much with this error. However, smaller distances or angular measurements can be questioned since the displacement of the anatomical point may completely alter the outcome. [3] Myers [12] asserts that posture, meaning standing or sitting still, does not exist because humans are never placed in stillness. In other words, there are always moving, shifting, balancing and adapting. This represents a further problem for picture-based postural analysis.

The aim of this chapter is to explore the literature in search for new scientific methods to assess posture using cameras and to discuss among both new and old which are the opitmal methods for scientific and clinical objectives.

Method

Search Methods

The Medline and Lilacs databases were consulted for relevant articles from 2003 to 2013 with the key words "posture" and "postural". Articles needed to be in English, Portuguese, French, Italian or Spanish. They also needed to have a description of an optical postural assessment.

Criteria for Inclusion and Exclusion

All articles that assessed posture with cameras in some way were considered. Reviews of postural assessment and articles that discussed posture in some manner that could assist with the discussion were also included. The quality of the article was not considered important if assessment was not its aim. Empirical research, letters to the editor and conference proceedings were excluded.

Study Selection

For all research articles identified during the search, the titles, keywords and abstracts of were read in order to confirm if they satisfied the inclusion criteria. Full text copies were obtained for analysis and data extraction for all articles that met the inclusion criteria. Preference was given to recent reviews on postural assessment and researches of new or unusual forms of evaluation. Older articles that showed the same information contained in the newer ones were also excluded.

Results

A total of 253 articles were found which assessed posture with cameras in some manner. It is clear that there are two types of technology currently being used: common cameras (2D photos) and motion capture (2D or 3D motion recording). This last modality was divided into motion capture with cameras and 3D sensors. From the 253 articles 49 were selected based on the inclusion and exclusion criteria.

Discussion

1. Pictures

Fortin et al. [13] stated that the measurement of body angle or distance by photography is the most promising technique to globally assess posture both in the sagittal (two sides) and frontal planes (anterior and posterior views) because photograph acquisition is cheap, fast, and easy. A number of studies have reported a reasonable correlation between radiographic measurements and the placement of markers. [14-16] A number of authors attempted to employ methods that would reduce the possibility of error in marking the bony landmarks and the correct placement of joint centers and axes. [17-21] Sacco et al. [22] studied the reliability of photographic assessment in relation to goniometry of the lower limbs. Twenty-six asymptomatic volunteers, with no differences greater than 1cm between the lower limbs were measured for the following data: ankle, knee flexion/ extension; rear foot; and the Q angle with a manual goniometer and digital photogrammetry. All of the results were similar, except for the Q angle. (Table 1) Based on these observations, it can be inferred that the software does not make much difference in the assessment, since all of the trace angles and distances were similar. The photograph is therefore quite similar to goniometry in terms of the assessment result. Smith et al. [23] compared angles of curvature of the spine in photographic and radiographic assessments, both in the standing position and lateral view of 766 teenagers. Since the main focus of this article was the association with pain, the authors did not directly correlate the two types of assessment. However, the classification of the alignment of the thorax, lumbar spine and pelvis was consistent between the two assessments, suggesting the use of photographs to avoid exposing the patients to radiation.

Table 1. Mean, standard deviation and p value for the goniometric, Corel Draw and SAPo assessments according to the work of Sacco et al. [22]

	Goniometry	Corel Draw	SAPo†	p
Ankle flexion/ extension	112.3 + 4.0	112.4 + 3.6	112.4 + 3.4	0.9991
Rearfoot angle	7.1 + 3.7	8.1 + 4.5	8.1 + 4.4	0.2159
Q angle	15.0 + 5.6	13.1 + 7.8	13.1 + 7.8	0.0068*
Knee flexion/ extension	184.0 + 4.7	181.7 + 4.1	181.6 + 4.3	0.4027

* Significant statistical difference. †Postural assessment software.

Iunes et al. [24] studied twenty-one volunteers, who were visually assessed by three experienced physiotherapists and then photographed with markers attached to the skin at various anatomical sites. The photographs, in turn, were analyzed by three other examiners. There was statistical concordance (Pearson correlation) between the examiners who used photogrammetry for all of the segments assessed. Inter-examiner correlations were low (r = .13 to .59) for 15 anatomical landmarks, moderate (r = .61 to .74) for 10, and high (r = .81 to .82) for four. Although it is interesting that this study reported few associations between the visual and photographic assessment, the study contained a methodological error. It concluded that the gold standard would be postural assessment by photograph. The problem with this assessment was the placement of the markers, which was unique. There was not a separate placement of markers for each physiotherapist, which would have assessed the real inter-correlation of the assessments.

Smith et al. [23] compared the alignment of the knee using photographs and radiographs and concluded that photos are a viable tool for this purpose. Engsberg et al. [25] compared skin surface markers and radiographic images and obtained some interesting findings. Skin surface markers representing bony landmarks were used to obtain distances and angles based on photographs. Placing metallic markers on C7 and S2 on 28 subjects before a biplanar radiograph, the authors measured the alignment of the marker and bone. Based on their findings, they suggested caution in interpreting results from only the surface marker. Fortin et al. [13] also discussed landmarks, reiterating that the ability to identify them can be a factor that affects the reliability of postural measures. Other factors that can compromise measurement are: the subject's physiological factors, such as a balance or sway problem during stance; the number of investigators; the landmarks chosen; and the way that postural body angles are calculated. [13-26] A review study has shown the most common landmarks used in science. [26] Table 2 displays these landmarks and Figure 1 shows examples of landmarks disposition.

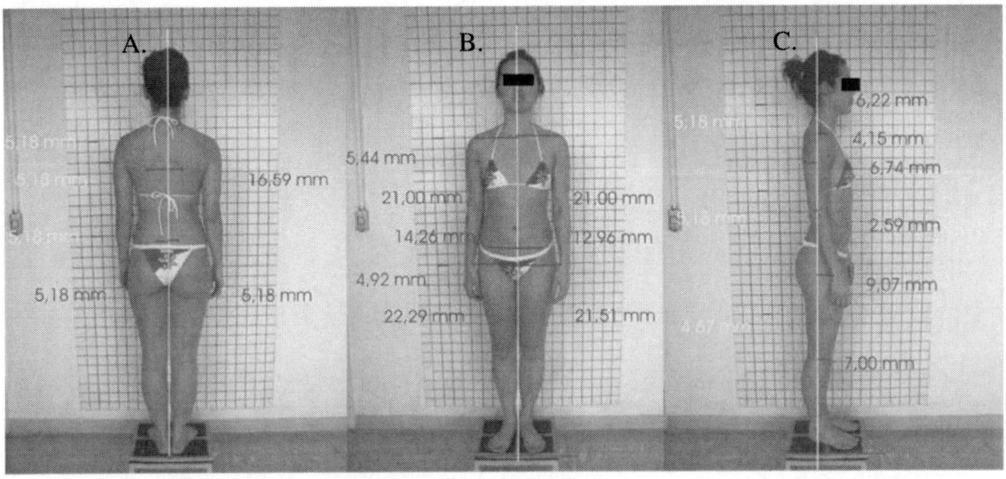

Figure 1. Examples of pictures of the frontal plane with ventral (A) and dorsal (B) incidence and sagittal plane (C). Landmarks were used in order to calculate postural deviations with the Corel Draw^{tm} software.

In table 1, it is notable that the spinous process of the seventh cervical vertebra is the most common anatomical landmark used. This point is relatively easy to find and can be used for many measurements of the spine, head and shoulders. Other spinous processes are commonly marked and used together to measure lordosis, kyphosis and scoliosis. However, care must be taken when counting the vertebrae and small Styrofoam balls, glued with double sided tape, should be used for the photos in lateral views, as reported in the work of Canales et al. [31] and Ferreira et al. [28]

The malleoli, fibular head, and greater trochanter of the femur are also widely used, probably because they are small bony prominences and are easy to access. Less common, but with the same localization logic are tibial tuberosity, chin protuberance, manubrium of the sternum and posterior calcaneal tuberosity.

The anterior superior and posterior superior iliac spines deserve special attention. These points are widely used, but are more difficult to find due to increased abdominal fat. Therefore, the scientific use of these points must be associated with a control parameter of this tissue, such as the body mass index or abdominal cirtometry in order to not to compromise the examination.

Table 2. List of anatomical landmarks used for postural assessment and the respective authors that utilized them

Anatomical landmarks	Authors
First metatarsophalangeal joint	Cobb et al. [27]
Midpoint between the second and third metatarsals	Ferreira et al. [28]
Navicular tuberosity	Cobb et al. [27]
Lateral malleolus	Miranda et al., [29] Saito et al., [30] Ferreira et al., [28] Canales et al. [31]
Medial malleolus	Miranda et al., [29] Cobb et al., [27], Ferreira et al. [28]
Posterior tuberosity of the calcaneus	Ferreira et al. [28]
Achilles tendon	Miranda et al., [29] Ferreira et al. [28]
The midpoint of the calcaneus	Miranda et al. [29]
Fibular head	Miranda et al., [29] Saito et al., [30] Canales et al. [31]
Tibial tuberosity	Ferreira et al. [28]
Joint line of the knee	Ferreira et al. [28]
Middle of the patella	Ferreira et al. [28]
Medial femoral condyle	Miranda et al. [29]
Greater trochanter of the femur	Miranda et al., [29] Saito et al., [30] Ferreira et al., [28] Canales et al. [31]
Anterior superior iliac spines	Miranda et al., [29] Saito et al., [30] Ferreira et al., [28] Canales et al., [31] Rosário et al. [3]
Posterior-superior iliac spines	Miranda et al., [29] Saito et al., [30] Ferreira et al., [28] Canales et al., [31] Rosário et al. [3]
Acromion	Miranda et al., [29] Thigpen et al., [33] Ferreira et al., [28] Canales et al., [31] Rosário et al. [3]
Coracoid process	Saito et al. [30]
Spinous processes:	
C7	Miranda et al., [29] Motta et al., [32] Saito et al., [30] Thigpen et al., [33] Ferreira et al., [28] Canales et al., [31] Engsberg et al., [25] Cuccia et al. [34]
T1	Miranda et al., [29] Claus et al. [35]
T3	Miranda et al., [29] Ferreira et al. [28]
T5	Claus et al. [35]
T6	Miranda et al. [29]
T7	Miranda et al. [29]
T10	Claus et al. [35]
T12	Miranda et al. [29]
L3	Miranda et al., [29] Claus et al. [35]
L5	Miranda et al. [29]
S2	Claus et al., [36] Engsberg et al. [25]
Inferior angle of the scapula	Miranda et al., [29] Saito et al., [30] Ferreira et al., [28] Rosário et al. [3]
Manubrium of the sternum	Motta et al., [32] Rosário et al., [3]
Chin protuberance	Motta et al. [32]
Tragus	Thigpen et al., [33] Ferreira et al., [28] Cuccia et al., [34] Rosário et al. [3]

The inferior angle of the scapula is also an interesting point, this can be easy to find, and is therefore less likely to lead to a methodological error. On the other hand, the acromion is a relatively large spot and requires a more specific point. The middle of the patella can easily generate errors, unless a tape measure is used to find the exact center. However, the patella can be dislocated in some people and lead to an assessment error despite its central point being perfectly located. The midpoint of the calcaneus is also a challenging location, since it is a large and irregular bone. The posterior calcaneal tuberosity however, seems to be a good substitute. The same problem exists with the femoral condyles, since it is very easy to produce errors because their size causes confusion in terms of their exact location.

The Achilles tendon is often used to evaluate the position of the rear foot. However, despite its clinical value, it is difficult to find a precise point for photographic measures through the tendon alone. If the point is poorly chosen, which can easily happen due to the length of the tendon, this may result in alterations of the measurement angles. The midpoint between the second and third metatarsal is also a vague point. The joint line of the knee may also not be an appropriate choice, because it is not a point but a region.

One exception to the points that are not a bony landmark is the tragus, which is a small and well-defined structure and as such is an easy location to find. In order to increase reliability, it is possible to use the intertragic notch at this region, which is even smaller and better defined.

The recommended landmarks suggested based on that review study were [26]: malleoli; posterior calcaneal tuberosity; fibular head; tibial tuberosity; greater trochanter of the femur; angle anterior and/or posterior to the lateral edge of the acromion; spinous processes (in particular C7); inferior angle of the scapula; manubrium of the sternum; chin protuberance; and the intertragic notch. Iliac spines, both posterior superior and anterior superior, should only be used in lean subjects.

From a positive perspective, pictures are appealing due their low cost and portability. It is even possible to find smart phones apps for postural analysis. On the other hand however, the chances of errors are significantly increased. Problems in pinpointing the landmarks and finding the correct position of the camera (small rotation can distort the picture) are the two major problems. They can be minimized with a more appropriate landmark choice [26] and fixed camera positioning, even though, there is a bias to consider in the method.

2. Motion Capture

3D posture analysis systems can assess posture in a quantitative fashion and may be more appropriate for understanding postural measures. However, these systems are not accessible to most clinicians treating persons with musculoskeletal or neurologic disorders. [13] Usually, 3D analysis reconstructs the body image based on a number of cameras (between 3 and 6) and reflective markers. [36]

Pazos et al. [37] employed the 3D trunk reconstruction for follow-up of scoliotic deformity. They noted that for over 20 years, many researchers have sought alternatives to radiographs for scoliosis assessment. However, correlation studies aimed at establishing a relationship between surface asymmetry and the spinal deformity underneath have been disappointing. This however may be as a consequence to the available technologies at that time. In the same study, the researchers discussed some reliability and accuracy issues in capturing 3D images, namely, postural sway and breathing patterns.

When 3D images are acquired with cameras and skin markers are used to evidence bony landmarks, the same problems encountered with pictures emerge. Gorton III et al. [38] also aimed to study scoliosis tested with a 3D scanner. Considering that surface topography is a noninvasive alternative to radiography for quantifying the body shape and does not use landmarks, this may be a promising solution. However, to the author's knowledge, the use if this method still requires validation and further study.

To summarize, 3D images have several advantages over pictures. It can assess the patient from different angles simultaneously, assess not just the posture, but also gait, and assess the dynamical part of the posture. This in turn satisfies the non-stillness postural concept. [12] However, besides the same landmarks problems observed in pictures, motion capture systems can be far more expensive and take much more space, being a more challenging system to transport. In contrast, a markerless system using a good algorithm can be an option to be tested for accuracy.

3. 3D Sensors

Camera-based systems suffer for passive vision problems such as brightness variation, shadows, perspective ambiguity (monocular), optical occlusions (stereo), correspondence search (stereo), intrinsic camera calibration and high computational costs. [39] Besides, existing laboratory-based 3D motion capture systems are of limited use for clinical environments. Both active (e.g., NDI) and passive (e.g., Vicon Motion Systems) video based systems are not easy to use in real-world applications due to complexity, bulk and space requirements. [40]

3D sensors have been increasingly investigated as solutions being able to overcome these drawbacks. [41] Actually, 3D sensors processes in-hardware processed dense depth maps at high frame rates and at every point in the scene, which reduces the overall computational workload.

3D sensors became popular after Microsoft® used their depth and image sensor named *Kinect*, usually found together with X-box 360 gaming console. It can detect movements and identify faces, allowing players to use only their own bodies as controls while playing games. Unlike previous systems of gesture or movement-based controls, this device does not require the player to wear any kind of accessory to track the player's movements. The *Kinect* follows users' movements by tracking and identifying their joints in three-dimensional space, which are obtained from the sensor data. However, the application of this equipment has been used to other purposes and connected to common computers. It uses infra-red light and a video camera creating a 3D map of the area in front of its cameras, [42] and uses an algorithm to automatically determine body landmarks, with a 3D camera based motion analysis system. [43] An RGB (a camera that delivers the three basic color components - red, green, and blue - on three different wires) gets a 2-dimensional color video of the scene for facial identification and for displaying images on the screen during play.

The depth component of the *Kinect* is made up of two sensors that are the basis for gesture recognition and skeleton tracking when working together: an infrared projector and a monochrome CMOS sensor. [44] The infrared light projector shines a grid of infrared light on the field of view. When the sensor receives back the rays from reflections of the light off of objects in the scene a depth map is created, which specifies the distance of the surfaces of objects from the viewpoint of the *Kinect.*

Dutta [45] compared the *Kinect* with a Vicon 3D motion capture system and reported that the former showed less precision than the 3D system. However, the author concluded that with a small amount of further development, *Kinect* may become a portable 3D motion capture system. Clarck et al. [46] stated that the *Kinect* can be compared to a 3D motion analysis system when assessing anatomical landmark position and angular displacement during common postural control tests. According to these authors, Microsoft® *Kinect* has both benefits and drawbacks when compared to a marker-based 3D camera system for 3D anatomical landmark assessment. The major benefits are the cost, portability and that it is markerless. Using markers presents several potential problems, [47] such as soft tissue artifact and setup time required to accurately position the markers over the respective anatomical landmarks, among others. On the other hand, data collection time can be greatly decreased when using a markerless system, with only the setup time being the calibration protocol and having the individual change into reasonably tight clothing to ensure data integrity. An additional benefit is that the anatomical landmark data is automatically determined in close to real-time by the machine-learning algorithm, and therefore the results can be provided to the patient almost immediately.

An important drawback of the *Kinect* may be the presence of proportional biases for some outcome measures. For example, Della Croce et al. [48] found that the data for the sternum derived from the *Kinect* tended to be higher than that derived from the 3D camera system. Thus, greater movement amplitude of the sternum would potentially increase the differences between the methods. Of course, this is based on the assumption that the systematic biases were due primarily to the *Kinect* and not the full-body plug-in-gait marker model the authors utilized. Although, it is well known that models using skin based markers are prone to systematic errors. Despite its lower precision, *Kinect* has the same advantages and disadvantages as a markerless 3D motion system with the additional. [45] The major challenge of motion recognition with *Kinect* is the noisiness and incompleteness of the tracked postures, suffering from similar problems with traditional optical motion capturer, such as occlusions and mixing up of tracked body parts. [49] Another limitation of the *Kinect* identified by Clark et al. [46] is the inability to assess internal/external joint rotations in the peripheral limbs because it does not possess the ability to accurately determine a non-joint center related orthogonal axis to the primary longitudinal axis.

The new generation of *Kinect* released in 2013 is using a new type of sensor named 'time of flight' (ToF), which functions in a similar way but it is more accurate. A ToF camera has a variety of advantages over alternative 3D scanning technology: It can measure 3D depth maps at video rate and thus lends itself for integration into a fast object scanner. It senses depth by emitting a pulse or modulated light signal (infrared) and measures its travel time, and therefore it does not interfere with the scene in the visual spectrum. Its core components are a CMOS chip (Complementary Metal-Oxide-Semiconductor) and an infrared light source which bears the potential for low cost production in big volumes.

ToF sensors are a technology that offers rich sensory information about a large part of the scene and, at the same time, enables a convenient, non-invasive system setup. These sensors provide dense depth measurements at every point in the scene at high frame rates. The range data provided allows easy segmentation of the human body and can also disambiguate poses that would otherwise have similar appearance and therefore confuse most monocular systems.

To summarize, 3D sensors such as Microsoft® *Kinect* are not yet the best solution, but they may be in a near future. They are inexpensive, portable and do not require markers to determine anatomical landmarks, and consequently may overcome the limitations associated with laboratory-based movement analysis systems. The drawback is that they still have a lack of precision. However, new technologies such as the ToF cameras are improving in accuracy.

Conclusion

Posture evaluation is a challenging task. All systems, namely: common cameras, motion capture systems and 3D sensors, using markers or not, have positive points and drawbacks. Researchers and clinicians should carefully choose their assessment equipment based on accuracy, pre-assessment preparation time, space available for equipment installation and cost. It is anticipated that future technologies tend to be more cost effective and more accurate.

References

[1] Liebenson C. Postural correction. *J Bodyw Move Ther.* 2008;12:318-9.

[2] Wallden M. The neutral spine principle. *J Bodyw Move Ther.* 2009;13:350-61.

[3] Rosário JLP, Nakashima IY, Rizopoulos K, Kostopoulos D, Marques AP. Improving posture: Comparing Segmental Stretch and Muscular Chains Therapy. *Clin Chiropractic.* 2012;15: 121-8.

[4] Missaoui B, Portero P, Bendaya S, Hanktie O, Thoumie P. Posture and equilibrium in orthopedic and rheumatologic diseases. *Neurophysiol Clin.* 2008;38:447-57.

[5] Suzuki H, Endo K, Mizuochi J, Kobayashi H, Tanaka H, Yamamoto K. Clasped position for measurement of sagittal spinal alignment. *Eur Spine J.* 2010;19:782-6.

[6] Berthonnaud E, Dimnet J, Hilmi R. Classification of pelvic and spinal postural patterns in upright position. Specific cases of scoliotic patients. *Computerized Medical Imaging and Graphics.* 2009; 33: 634-43.

[7] Steffen J-S, Obeid I, Aurouer N, Hauger O, Vital J-M, Dubousset J, Skalli W. 3D postural balance with regard to gravity line: an evaluation in the transversal plane on 93 patients and 23 asymptomatic volunteers. *Eur Spine J.* 2010;19:760-7.

[8] Viguier M, Dupui P, Montoya R. Posture analysis on young women before and after 60 days of -6 degrees head down bed rest (Wise 2005). *Gait Posture.* 2009;29:188-93.

[9] Bonde-Petersen F. A simple force platform, *Euro Appl Physiology.* 1975; 34:51-4.

[10] Nashner LM. Vestibular postural control model. *Kybernetik.* 1972; 10:106-10.

[11] Nashner LM, McCollum G. The organization of human postural movements: A formal basisand experimental synthesis. *Brain Behav.* 1985; 8:135-72.

[12] Myers T. Acture! Posture in action. *Massage & Bodywork Magazine.* 2006.

[13] Fortin C, Feldman DE, Cheriet F, Labelle H. Clinical methods for quantifying body segment posture: a literature review. *Disabil Rehabil.* 2011;33:367-83.

[14] Hunt MA, Birmingham TB, Jenkyn TR, Giffin JR, Jones IC. Measures of frontal plane lower limb alignment obtained from static radiographs and dynamic gait analysis. *Gait Posture.* 2008;27:635-40.
[15] Mundermann A, Dyrby CO, Andriacchi TP. A comparison of measuring mechanical axis alignment using three-dimensional position capture with skin markers and radiographic measurements in patients with bilateral medial compartment knee osteoarthritis. *Knee.* 2008;15:480-5.
[16] Vanwanseele B, Parker D, Coolican M. Frontal knee alignment: three-dimensional marker positions and clinical assessment. *Clin Orthop Relat Res.* 2009;467:504-9.
[17] Bell AL, Pedersen DR, Brand RA. A comparison of the accuracy of several hip center location prediction methods. *J Biomech.* 1990;23:617-21.
[18] Camomilla V, Cereatti A, Vannozzi G, Cappozzo A. An optimized protocol for hip joint centre determination using the functional method. *J Biomech.* 2006;39:1096-106.
[19] Ehrig RM, Taylor WR, Duda GN, Heller MO. A survey of formal methods for determining the centre of rotation of ball joints. *J Biomech.* 2006;39:2798-809.
[20] Ehrig RM, Taylor WR, Duda GN, Heller MO. A survey of formal methods for determining functional joint axes. *J Biomech.* 2007;40:2150-7.
[21] Taylor WR, Ehrig RM, Duda GN, Schell H, Klein P, Heller MO. On the influence of soft tissue coverage in the determination of bone kinematics using skin markers. *J Orthop Res.* 2005;23:726-34.
[22] Sacco, ICN, Alibert S, Queiroz BWC, Pripas D, Kieling I, Kimura AA, et al. Confiabilidade da fotogrametria em relação a goniometria para avaliação postural de membros inferiores. *Rev Bras Fisioter.* 2007;11:411-17.
[23] Smith A, O'Sullivan PB, Straker L. Classification of sagittal thoraco-lumbo-pelvic alignment of the adolescent spine in standing and its relationship to low back pain. *Spine.* 2008; 33:2101-7.
[24] Iunes DH1, Bevilaqua-Grossi D2, Oliveira AS2, Castro FA3, Salgado HS Análise comparativa entre avaliação postural visual e por fotogrametria computadorizada. *Rev Bras Fisioter, São Carlos.* 2009;13:308-11.
[25] Engsberg JR, Lenke LG, Bridwell KH, Uhrich ML, Trout CM. Relationships between spinal landmarks and skin surface markers. *J Appl Biomech.* 2008;24:94-7.
[26] Rosário JLP. Photographic analysis of human posture: a literature review. *J Bodyw Move Ther.* In press.
[27] Cobb SC, James CR, Hjertstedt M, Kruk J. A digital photographic measurement method for quantifying foot posture: Validity, reliability, and descriptive data. *J Athletic Training.* 2011;46:20-30.
[28] Ferreira EAG, Duarte M, Maldonado EP, Burke TN, Marques AP. Postural assessment software (PAS/SAPO): validation and reliability. *Clinics.* 2010;65:675-81.
[29] Miranda R, Schor E, Girão MJBC. Avaliação postural em mulheres com dor pélvica crônica. *Rev Bras Ginecol Obstet.* 2009; 31:353-60.
[30] Saito ET, Akashi PM, Sacco Ide C. Global body posture evaluation in patients with temporomandibular joint disorder. *Clinics.* 2009;64:35-9.
[31] Canales JZ, Cordás TA, Fiquer JT, Cavalcante AF, Moreno RA. Posture and body image in individuals with major depressive disorder: a controlled study. *Rev Bras Psiquiatr.* 2010;32:375-80.

[32] Motta JL, Martins MD, Fernandes KPS, Mesquita-Ferrari RA, Biasotto-Gonzalez DA, Bussadori SK. Craniocervical posture and bruxism in children. *Physiother Res Int.* 2011;16:57-61.

[33] Thigpen CA, Padua DA, Michener LA, Guskiewicz K, Giuliani C, Keener JD, Stergiou N. Head and shoulder posture affect scapular mechanics and muscle activity in overhead tasks. *J Electromyogr Kinesiol.* 2010;20:701-9.

[34] Cuccia AM, Carola C. The measurement of craniocervical posture: A simple method to evaluate head position. *Int J Pediatr Otorhinolaryngol.* 2009;73:1732-6.

[35] Claus AP, Hides JA, Lorimer MG, Hodges PW. Is 'ideal' sitting posture real? : Measurement of spinal curves in four sitting postures. *Manual Therapy.* 2008;14:404-408.

[36] Sawacha Z, Carraro E, Del Din S, Guiotto A, Bonaldo L, Punzi L, Cobelli C, Masiero S. Biomechanical assessment of balance and posture in subjects with ankylosing spondylitis. *J Neuroeng Rehabil.* 2012;9:63.

[37] Pazos V, Cheriet F, Song L, Labelle H, Dansereau J. Accuracy assessment of human trunk surface 3D reconstructions from an optical digitising system. *Med Biol Eng Comput.* 2005;43:11-5.

[38] Gorton GE 3rd, Young ML, Masso PD. Accuracy, reliability, and validity of a 3-dimensional scanner for assessing torso shape in idiopathic scoliosis. *Spine* (Phila Pa 1976). 2012;37:957-65.

[39] Hartley RI, Zisserman A. Multiple view geometry in computer vision (2nd ed.). Cambridge: University Press; 2004.

[40] Best R, Begg R. Overview of movement analysis and gait geatures. In: Begg, R., Palaniswami, M. (Eds.), Computational intelligence for movement sciences: Neural networks and other emerging techniques. Pennsylvania: Idea Group Publishing, Hershey; 2006.

[41] Ganapathi V, Plagemann C, Thrun, Koller D. Real time motion capture using a single time-of-flight camera. In Proc of CVPR. 2010;755-62.

[42] Menna F, Remondino F, Battisti R, Nocerino E. Geometric investigation of a gaming active device. Proceedings of SPIE – The International Society for Optical Engineering 2011;8085:80850G.

[43] Shotton J, Fitzgibbon A, Cook M, Sharp T, Finocchio M, Moore R, et al. Real-time human pose recognition in parts from single depth images. In: Proceedings of the IEEE computer society conference on computer vision and pattern recognition. 2011;1297-304.

[44] Orendurff MS, Segal AD, Klute GK, Berge JS, Rohr ES, Kadel NJ. The effect of walking speed on center of mass displacement. *J Rehab Res Dev.* 2004;41:829-34.

[45] Dutta T. Evaluation of the Kinect™ sensor for 3-D kinematic measurement in the workplace. *Appl Ergon.* 2012;43:645-9.

[46] Clark RA, Pua YH, Fortin K, Ritchie C, Webster KE, Denehy L, Bryant AL. Validity of the Microsoft Kinect for assessment of postural control. *Gait Posture.* 2012; 36:372-7.

[47] Mündermann L, Corazza S, Andriacchi TP. The evolution of methods for the capture of human movement leading to markerless motion capture for biomechanical applications. *J Neuroengineering Rehab* 2006;3:6.

[48] Della Croce U, Leardini A, Chiari L, Cappozzo A. Human movement analysis using stereophotogrammetry. Part 4: assessment of anatomical landmark misplacement and its effects on joint kinematics. *Gait Posture* 2005;21:226-37.

[49] Shum HP, Ho ES, Jiang Y, Takagi S. Real-time posture reconstruction for Microsoft Kinect. *IEEE Trans Cybern.* 2013;43:1357-69.

In: Posture: Types, Exercises and Health Effects
Editor: Sarah A. Curran
ISBN: 978-1-63117-252-6

Chapter 2

The Clinical Usefulness of Head Posture Assessment for Patients with Neck Pain

Anabela G. Silva,** *PhD, MSc, BSc., David Punt, PhD and Mark I. Johnson, PhD, BSc, PGCertHE

[1]School of Health Sciences, University of Aveiro, Agras do Castro, Campus Universitário de Santiago, Aveiro, Portugal
[2]School of Sport, Exercise and Rehabilitation, University of Birmingham, United Kingdom
[3]Faculty of Health and Social Sciences, Leeds Metropolitan University, Portland Building, City Campus, United Kingdom

Abstract

Head posture assessment has long been advocated as an important part of the examination for patients with neck pain. However, for a comprehensive understanding of the clinical usefulness of head posture assessment for patients with neck pain there is a need to discuss a set of topics, including: i) head posture definition, ii) the biomechanical and physiological implications of head posture changes, iii) whether there are differences in head posture between patients with neck pain and asymptomatic individuals, iv) what are the appropriate measurement procedures, v) what exercises are effective for head posture correction, and, vi) what are the benefits of head posture correction for patients with neck pain. We discussed these points in light of the most recent research, including our own findings on the subject.

Keywords: Head posture, neck pain, examination, standardization, measurement

* Corresponding author: Email: asilva@ua.pt

Introduction

There is a long held belief that posture influences health and wellbeing, dating back to the ancient Greeks who reflected their understanding of ideal posture in their statues. In 1927, Schwartz [1] reviewed literature on posture and suggested a list of consequences of not having ideal posture that included disorders of the cardiac, respiratory, digestive, neurological and musculoskeletal systems. In 1947, the American Academy of Orthopedic Surgery defined good posture as a state of optimal balance that protects the muscular and skeletal structures against injury and deformity and promotes maximum body efficiency. [2]

The assessment of head posture (HP) is recommended as part of the examination of patients with neck pain. [3-5] Clinicians assess HP by judging the alignment of the head against imaginary lines of reference crossing specific anatomical points. [6] It is believed that the examination of HP for patients with neck pain can be useful in aiding diagnosis, determining subsequent treatment strategies and in monitoring progress of the patient. [6-8]

Neck pain is a highly prevalent disorder in the general population, with a lifetime mean point prevalence of 48.5%. [9] For most patients, the aetiology of neck pain is difficult to establish and it is claimed that assessment of HP may provide useful information about the pathophysiological drivers of neck pain. [6] The usefulness of HP assessment depends on two assumptions. Firstly, there are differences in the measurements of various aspects of HP between patients with neck pain and asymptomatic individuals. Secondly, these differences can be measured consistently (reliability) and clinical inferences can be based on the measurements obtained (validity).

The purpose of this chapter is to discuss the clinical usefulness of HP assessment for patients with neck pain. This will be achieved by firstly establishing operational definitions for 'ideal' HP, 'normal' HP and 'abnormal' HP. Protocols and procedures used to assess HP will then be discussed, evaluating research on relationships between HP and neck pain and reviewing the effectiveness of procedures used to correct HP in clinical practice.

Definition of Head Posture

Posture refers to the alignment of body parts in relation to each other. [8,10] Hence, HP refers to the alignment of the head in relation to the rest of body and is usually characterised by four components in the three anatomical planes of reference. [2,8] These components are conveyed in relation to the neck respective movement as [11]:

- Forward HP or retraction (with forward HP sometimes referred to as protraction) in the sagittal plane;
- Head extension or flexion, in the sagittal plane;
- Left or right side-flexion, in the coronal plane;
- Left or right head rotation, in the transverse plane.

In clinical practice, HP is judged by observation against imaginary lines of reference crossing specific anatomical points and the deviation from these lines characterized using qualitative descriptors such as mild, moderate or severe. It is recommended that forward HP and head extension are assessed against a line of reference passing through the external auditory meatus, the bodies of most cervical vertebrae and the mid-shoulder. Side-flexion and head rotation are assessed against a line dividing the body into two symmetrical halves passing through the cervical spine and mid-skull, with asymmetry suggestive of abnormality. [2]

Forward HP is the most common deviation of HP described in the literature [12] and has been defined as the protrusion or projection of the head in the sagittal plane so that the head is placed anterior to the line of reference representing ideal HP [2,13] (Figure 1). According to Harrison et al., [14] forward HP can occur due to an anterior translation of the head, flexion of the lower cervical spine or both. It is also claimed that forward HP involves a combination of lower cervical flexion and upper cervical extension, being present in association with head extension. [13,15] Despite forward HP often being used synonymously with deviation from ideal HP, there is evidence that it is normal for asymptomatic individuals to have some degree of forward HP when compared to the ideal line of reference [16,17]. The other components of HP are less often discussed and measured in research. However, data suggest that it is normal to have some degree of head extension and no side flexion or rotation. [12,18]

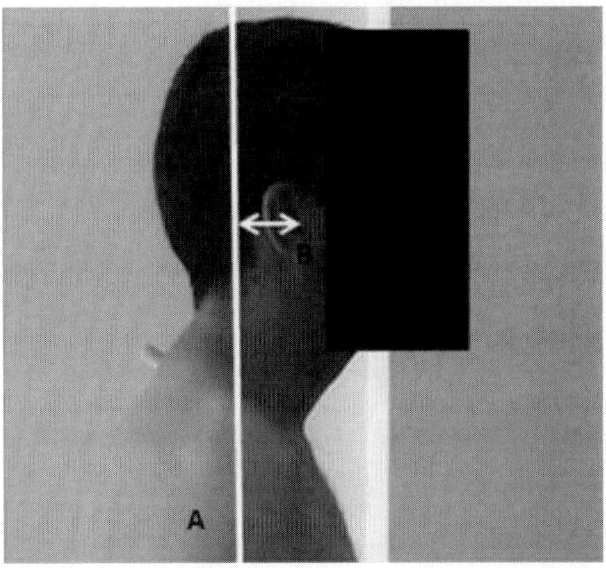

Figure 1. The white vertical line represents the line of gravity used to define ideal HP (A). This line should pass through the meatus of the ear. Thus, this individual presents with some degree of forward HP as the meatus of the ear lies ahead of the line (B - represents the distance between the meatus and the line).

In research, HP is characterized by measuring i) linear distances between two anatomical landmarks or ii) between one anatomical landmark and an external reference or iii) angles between anatomical landmarks as surrogates for HP. One surrogate for forward HP is the angle between the 7th cervical vertebra (C7), the tragus of the ear and the veridical horizontal (C7-tragus-horizontal). Decreasing values are indicative of a more forward HP. [19] Forward HP is also characterized as the distance between the tragus and an external reference such as a plumb line or a wall. [14] Head extension is usually characterized as the angle between the tragus, the canthus of the eye and the horizontal (tragus-eye- horizontal) and increasing values are indicative of a more extended head. Side-flexion is usually characterized as the angle between the inferior margins of both ears and the horizontal (right ear-left ear-horizontal) with 0° indicating perfect symmetry. [19] Rotation has been measured as the displacement of the head in the Z axis, [18] with 0 indicating no rotation, a positive value indicating rotation to the left and a negative value rotation to the right.

Ideal, Normal and Abnormal Head Posture

Different terms are used to characterize HP in the literature. These include 'ideal' HP, 'normal' HP and 'abnormal' HP. It is necessary to clarify these terms as, although they are used as synonymous, they have different meanings. A variety of definitions for 'ideal' HP exist. The definition of Kendall et al. [2, p.61] is the most commonly cited in the literature as "one in which the head is in a well-balanced position and maintained with minimum muscle effort. In side view, the line of gravity used as a reference coincides with the lobe of the ear and the neck presents the normal anterior curve. In posterior view, the line of reference coincides with the midline of the head and the cervical spinous processes. The head is not tilted upward or downward, and it is not tilted sideways or rotated. The chin is not retracted". However, research seems to suggest that this ideal HP is not common in a healthy population as, for example, the lobe of the ear is generally in front of the line of reference, [17] suggesting that ideal HP is mainly a theoretical concept.

In healthcare, normality is defined as a range of values that represent what is of high prevalence in a healthy (asymptomatic) population. [20] Therefore, normal HP corresponds to what is common in asymptomatic individuals. In studies on posture, normal range has been determined using one standard deviation from the population mean and also two standard deviations from the population mean. [21,22]

'Abnormal' posture would be postural values that fall outside of these scale ends. Thus, there is no consensus on normal and abnormal values for various measures of posture. A perhaps more useful, but also more challenging definition of what might be considered normal is provide by the Association of Faculties of Medicine of Canada Public Health in which normal is defined in terms of a range of scores above or below which treatment would be beneficial. [23] In terms of HP, this suggests the need to determine the limits of the range above or below which correcting HP might have an immediate or future benefit on neck pain. This will be further discussed on a specific section on the benefits of HP correction for patients with neck pain.

In this chapter, the following definitions will be used:

- 'Ideal' HP which corresponds to the anatomical references in the head being coincident with the line of reference as described by Kendal et al. [2]
- 'Normal' HP which is what is common in asymptomatic individuals and measurement values fall within one standard deviation from the population mean;
- 'Abnormal' HP corresponds to what falls outside one standard deviation from the population mean.

Characterization of Head Posture in Research Studies

A summary of research studies that have attempted to measure aspects of HP and characterized them as normal and abnormal using a range of values within one standard deviation of the mean are provided in Table 1. Three studies with sample sizes over 100 were found. [24-26] Hanten et al. [25] measured the perpendicular distance between a wall and a mark placed below the eye on the zygomatic arc using a metric ruler to characterize forward HP. Dalton and Coutts [24] and Raine and Twomey [26] used the angle between C7, the tragus of the ear and the horizontal as a surrogate of forward HP. This is also called the cranio-cervical angle and is the surrogate measure of forward HP most often used in research. [27] Decreasing values are indicative of a more forward HP.

In addition, Raine and Twomey [26] used the angle between the tragus of the ear, the canthus of the eye and the horizontal as a surrogate of head extension/flexion (positive values indicate extension and negative values indicate flexion) and the angle between the ears and the horizontal as a surrogate of head right and left side-flexion (negative and positive signs indicate flexion to opposite sides). Increasing values of the angle between the tragus of the ear, the canthus of the eye and the horizontal are indicative of a more extended head. Rotation was not measured in these studies.

Age and sex are generally considered as factors that affect normal ranges for posture. However, no differences were found between men and women or between age ranges for head extension/flexion and side-flexion in any of the studies. All studies agree that forward HP increases with age, but results are contradictory for sex as Dalton and Coutts [24] showed women to have a more forward HP than men, Hanten et al. [25] showed men to have more forward HP than women and Raine and Twomey [26] found no significant differences between men and women.

Biomechanical and Physiological Implications of Head Posture Deviations

Deviations from normal HP affect the mechanical arrangement of related body parts and are claimed to result in [2,13,28,29, 31, 33,59]:

- Changes in lever arms (either a decrease in the muscle moment arm or an increase in the resistive moment arm);
- An increase in muscle activity;
- An increase in the internal forces acting on the head and neck;
- Muscle ischemia and the release of compounds that stimulate nociceptors and lead to pain;
- Subcranial and upper thoracic restrictions and hypermobility of the midcervical spine;
- Impaired proprioception.

Table 1. Normative data from studies that measured the C7-tragus-horizontal angle (indicative of forward HP), the tragus-eye-horizontal angle (indicative of head extension) and the ear-left ear-horizontal (indicative of side-flexion) (♀ = male, ♂ = female)

Author	Aspect of HP measured	Instructions to participants	Position	N	Age Range (years)	Results (mean±SD)	Normal range (mean±1SD)
Hanten et al. [25]	Forward HP (measured as the distance between a wall and the zygomatic arc) (cm)	Assume a relaxed natural posture	Standing	218 (106♀,112♂)	20-30	♀=18.5±1.6, ♂=21.8±1.9	♀=16.9-20.1 ♂=19.9-23.7
					31-40	♀=19.5±2.4, ♂=22.5±1.7	♀=17.1-21.9 ♂=20.8-24.2
					41-50	♀=19.0±2.0, ♂=22.8±2.0	♀=17.0-21.0 ♂=10.8-24.8
					51-60	♀=19.4±1.9, ♂=22.5±2.2	♀=17.5-21.3 ♂=20.3-24.7
Dalton and Coutts [24]	Forward HP (measured as the angle C7-tragus-horizontal) (degrees)	Self-balanced position (participants tilted their head forwards and backwards until they felt that a natural HP was reached)	Seated	190 (93♀,97♂)	24-34	♀=49.5±3.5 ♂=50.6±4.0	♀=46.0-53 ♂=46.6-54.6
					35-44	♀=48.6±4.6, ♂=48.9±4.1	♀=44.0-53.2 ♂=44.8-53.0
					45-54	♀=46.8±4.3, ♂=49.3°±4.9	♀=42.5-51.1 ♂=44.4-54.2
					55-66	♀=42.0±5.0 ♂=46.4±3.8	♀=37.0-47.0 ♂=42.6-50.2
Raine and Towmey [26]	Forward HP (measured as the angle C7-tragus-horizontal) (degrees)	Look forward	Standing	167 (87♀,78♂)	17 - 29	♂=52.2±5.2 ♀=51.9±4.4	♀=37.0-57.4 ♂=4.5-56.3
					30 – 54	♂=47.6±5.9 ♀=50.8±4.8	♀=41.7-53.5 ♂=46.0-55.6
					≥ 55	♂=44.0±7.9 ♀=46.8±6.2	♀=36.1-51.9 ♂=40.6-53.0
	Head extension (measured as the angle tragus-eye-horizontal) (degrees)	Look forward	Standing	166 (86♀,78♂)	17 - 29	♂=5.9±5.5 ♀=5.8±5.7	♀=0.4-11.4 ♂=0.1-11.5
					30 – 54	♂=7.5±5.6 ♀=9.9±4.8	♀=1.9-13.1 ♂=5.1-14.7
					≥ 55	♂=7.8±7.1 ♀=11.2±6.9	♀=0.7-14.9 ♂=4.3-18.8
	Head side-flexion (measured as the angle right ear-left ear-horizontal) (degrees)	Look forward	Standing	160 (84♀,76♂)	17 - 29	♂=0.0±2.0 ♀=0.0±2.1	♀=-2.0-2.0 ♂=-2.1-2.1
					30 – 54	♂=0.4±3.0 ♀=0.2±2.8	♀=-2.6-3.4 ♂=-2.6-3.0
					≥ 55	♂=0.4±2.7 ♀=0.2±2.9	♀=-2.3-3.1 ♂=-2.7-3.1

Table 2. Studies comparing HP between participants with neck pain and asymptomatic participants

Study	Participants with NP	Asymptomatic participants	HP measurements	Position.	Results	Reviewer's conclusion
Harrisoon et al. [14]	n=10; age:23-43; sex:9♀,1♂	n=41; age:20-45; sex:30♀,11♂	i) C7-tragus-horizontal angle; ii) Tragus-eye-horizontal angle; iii) ⊥Distance between a plumb line through the malleolus and the tragus.	Static standing.	(in degrees) i) NP=49.4±4.2; AS=49.3± 7 ii) NP=21.6±6.4; AS=18.8±4.2 iii) (in cm) NP=6.7±1.9; AS=8.1±2.6	No differences between groups
Hanten et al. [97]	n=42; age:20-60; sex:NR.	n=42; matched to participants with NP for age and sex.	Distance between a wall and a point 3 cm below the corner of the eye.	i) Static standing; ii) Static sitting looking into their own eyes in a mirror.	(in cm) i) NP:♀=19.7±2.5; ♂=22.7±2.8 AS: ♀=19.1±2.3; ♂=22.2±1.6 ii) NP: ♀=47.1±23.7 ♂=50.0±20.0 AS: ♀=43.4±14.3; ♂=39.5±11.8	No differences between groups
Visscher et al. [95]	n=10; age:NR; sex:NR.	n=45; age:NR; sex:NR	C7-tragus-horizontal angle.	Static standing and sitting (mean values of both conditions used)	(in degrees) NP=51.1±6.5 AS=52.3±4.5	No differences between groups
Szeto et al. [98]	n=8; age:32.2±5.6 sex:8♀	n=8; age:30.7±6.6 sex:8♀	i) Forehead-tragus-vertical angle; ii) C7-tragus-vertical angle.	Static sitting; Sitting while performing a computer task	(in degrees) Static sitting i) NP= 55.4±5.5; AS=54.8±5.5 ii) NP=54.5±9.4; AS=47.0±4.7 Sitting working i) NP=60.6±6.1; AS=57.1±6.2 ii) NP=59.3±10.0; AS=52.5±7.1	No differences between groups
Lee et al. [99]	n=14; age:29.5±7.3;sex:7♀,7♂	n=26; age:26.6±6.7 sex:13♀,13♂	Distance between C7 and the posterior arm of the CROM.	Sitting: i) Upright; ii) Comfortable.	(in cm) i) NP=17.9±1.4; AS=18.7±1.3 ii) NP=20.0±1.6; AS=19.3±1.9	No differences between groups
Nilsson and Soderlund [100]	n=27; age:20-54; sex:13♀,14♂	n=40; age:20-52 sex:20♀,20♂	Angle between posterior aspect of the shoulder, meatus of the ear and vertical.	Static standing.	(in degrees) NP=44.6±4.9; AS=40.2±4.8	Differences between groups

Table 2. (Continued)

Study	Participants with NP	Asymptomatic participants	HP measurements	Position.	Results	Reviewer's conclusion
Szeto et al. [18]	n=21; age:20-45; sex:NR	n=17; age:NR; ± 5 years from participants with NP sex:17♀	i)Flexion/extension; ii) Side-flexion; iii) Rotation Measured as displacement of the head in the x, y, z axes.	Sitting while performing a writing task in the computer.	(in degrees) Median: i) NP=67.6±10.8; AS=63.7±12.9 ii) NP=-1.1±1.8; AS=-2.7±2.2 iii) NP=1.8±2.4; AS=4.2±2.5 Range i) NP=6.77±3.8; AS=4.53±2.1 ii) NP=3.56±2.1; AS=2.99±1.5 iii) NP=4.38±1.6; AS=3.41±1.3	Differences between groups for range values but not for median values
Edmondston et al. [68]	n=21; age:20-45; sex:NR	n=22; age:20-45; sex:NR.	i) T4-C7-tragus, ii) Tragus-eye-vertical.	Sitting: -Perceived good HP; -Habitual HP.	(in degrees) Perceived good HP: i) NP=153.6±5.9; AS=151.5±4. ii) NP=59.8±7.0; AS=64.4±7.8 Habitual HP: i) NP=158.0±5.8; AS=157.0±6.2 ii) NP=64.8±5.4; AS=68.0±7.3	i)No differences between groups ii) Differences between groups for perceived good HP but not for habitual HP
Falla et al. [94]	n=58; age:37.9±10.2 sex:58♀	n=10; age:35.0±4.6 sex:NR.	C7-tragus-horizontal angle.	Sitting for 10 minute playing a computer game.	(in degrees) Mean change during the task: NP:4.4±4.o; AS: 2.2±1.6	Differences between groups
Yip et al. [96]	n=62; age:39.9±10.8 sex:40♀, 22♂	n=52; age:42.3±11.2 sex:36♀+16♂.	C7-tragus-horizontal angle.	Static standing.	(in degrees) NP=49.9±6.1 AS=55.0±2.9	Differences between groups
Arvidsson et al. [101]	n=13; age:38 (27-55); sex:13♀	n=11; age:35 (25-51) sex:11♀	Flexion/extension measured with an inclinometer.	Sitting while performing a working task (air traffic controllers).	(in degrees) Median values: NP=8.0±6.0; AS=6.0±3.0	No differences between groups

Study	Participants with NP	Asymptomatic participants	HP measurements	Position.	Results	Reviewer's conclusion
Lau et al. [93]	n=26 age:36.9±10.0 sex:15♀,11♂	n=27; age:31.9±7.6 sex:15♀,12♂	C7-tragus-horizontal angle.	Static standing.	(in degrees) NP=43.9º±3.0; AS=50.6º±2.1	Differences between groups
Straker et al. [102]	n=682 age:14±? sex:?	n=788 age:14±? sex:?	) Tragus-eye-vertical; ii) C7-tragus-vertical; iii) T12-C7-tragus.		(in degrees) Looking ahead: i) NP=71.0±10.0; AS=72.0±10.0 ii) NP=53.0±9.0; AS=52.0±8.0 iii) NP=148.0±8.0; AS=150.0±8.0 looking down at their lap: i) NP=105.0±14.0; AS=106.0±13 ii) NP=69.0±10.0; AS=70.0±11.0 iii) NP=134.0±9.0; AS=134.0±9.0	No differences between groups

NR – not reported; ⊥ - perpendicular; CROM – cervical range of motion device; WAD – whiplash-associated NP classified according to the Quebec Task Force; NRS – numeric rating scale; NP – participants with NP, AS – asymptomatic participants. (♀ = male, ♂ = female).

Table 3. Studies investigating the effect of correcting HP in patients with neck pain

Author	HP measurement	Outcome	Participants	Type of exercise	Results
Abdulwahab and Sabbahi [107]	C7-tragus-horizontal angle	i) H reflex amplitude ii) Pain intensity (VAS)	Case group: 13 participants with complaints of neck, shoulder and arm pain and hand paraesthesia;	Measurements were taken before and after 20 minutes of reading and 20 repetitions of head retractions.	i) before reading=0.81±0.47 after reading=0.68±0.39, p>0.05; after retractions=1.01±0.49, p<0.01. ii) before reading=4.2±1.3 after reading=5.6±1.4; p<0.05 after retractions=1.5±1.3, p<0.01.
			Control group: symptomatic participants		i) before reading=1.11±0.65 after reading=1.06±0.59, after retractions=1.13±0.59, p>0.05. No between group differences were found.
Falla et al. [94]	C7-tragus-horizontal angle	Pain intensity (VAS)	Case group: 29 subjects with NP complaints of more than 3 months	6 week training of the deep craniocervical flexor muscles.	Mean difference for posture= 2.1°, p<0.01 Mean difference for pain=-0.9±2.3, p<0.05.
			Control group: 29 subjects with NP complaints of more than 3 months	6 week endurance-strength training for the neck flexor muscles	Mean difference for posture== 0.01°, p>0.05 Mean difference for pain = -1.1±2.8, p<0.05 Significant differences are within group differences before and after treatment
Diab and Moustafa [105]	C7-tragus-horizontal angle	i) C6 dermatomal somatosensory-evoked potentials ii) C7 dermatomal somatosensory-evoked potentials iii) pain intensity (VAS)	Case group: 48 patients with cerical spondylotic radiculopathy	10 week of superficial heat, continuous ultrasound, exercise (strengthening of the deep cervical flexors and shoulder retractors and stretching of cervical extensors and pectoralis)	Posture Before treatment=34.3±4.1 Immediately after treatment=41.07±2.9 At 6 months= 39.5±3.3 C6 potentials Before treatment= 0.41±0.18 Immediately after treatment=0.82±0.13 At 6 months= 0.79±0.12 C7 potentials Before treatment= 0.31±0.17 Immediately after treatment=0.82±0.13 At 6 months= 0.59±0.12

Author	HP measurement	Outcome	Participants	Type of exercise	Results
			Control group: 48 patients with cervical spondylotic radiculopathy	10 week of superficial heat, continuous ultrasound	Posture Before treatment=35.8±6.3 Immediately after treatment=34.8±3.3 At 6 months=34.5±3.4 C6 potentials Before treatment=0.42±0.21 Immediately after treatment=0.56±0.19 At 6 months=0.41±0.17 C7 potentials Before treatment=0.32±0.21 Immediately after treatment=0.43±0.19 At 6 months= 0.28±0.18 All differences between case and control groups were significant at $p>0.05$

All these potential changes allegedly associated with abnormal HP are believed to lead to pathology and pain. [2,13,28,29] Whether existing research supports the assumption that HP deviations from the norm result in biomechanical and physiological changes is addressed below.

Head Posture and Muscles Moment Arms

Evidence exists that supports the claim that HP deviations affect the length of the muscle moment arm, defined as the distance from the muscle's line of action to the joint's centre of rotation. [30] Vasavada et al. [31] developed a biomechanical model of the head and neck to evaluate changes on the length of neck muscle moment arms in different HPs. They reported the length of the muscle moment arms of most extensor muscles, including the suboccipital muscles, to decrease by 1 cm from extended to flexed HPs. Przybyla et al. [32] measured the length of muscles moment arms in vivo using magnetic resonance imaging scans and also reported a decrease in the length of extensors moment arms when the head was in flexion compared to when it was in extension. This decrease in the length of the neck extensor muscles moment arm results in a mechanical disadvantage during flexion, i.e., an increase in muscle activity will be needed to counteract the head's tendency to tip forward.

Head Posture, Muscle Activity and Force/Strength

When comparing muscle activity in a simulated ideal HP against participants' usual HP and a simulated forward HP, studies show conflicting results. McLean [16] compared muscle activity in 18 asymptomatic participants and found increased muscle activity when individuals stood in an ideal posture compared to their habitual forward HP. In contrast, Enwemeka et al. [33] measured trapezius muscle activity in 10 asymptomatic participants and reported a statistically significant increase in forward HP when compared to habitual or ideal HP. However, studies fail to report whether simulated HPs are mid-range postures or end-of-range postures and this may influence the muscular activity reported. Studies on other parts of the body have shown that muscles play a more relevant role in the stabilization of mid-range postures, while end-of-range postures seem to be mainly stabilized by inert structures such as the ligaments and joint capsules. [34,35]

In addition, research also seems to support an association between HP, muscle activity and muscle ability to produce force. Kumar et al. [36] studied the electromyography (EMG) and isometric force of sternocleidomastoid, splenii and trapezii muscles of 40 asymptomatic participants. They found that globally, EMG activity was approximately 66% higher in head flexion than in extension, while force output was approximately 30% less in flexion than in extension. This indicates that more muscle activity was required in flexion than extension to generate the same force. Similar results were reported by Mayoux-Benhamou and Revel [37] who investigated the influence of HP on muscular efficiency (calculated as the ratio between muscular activation and the force produced). They found that less extensor muscle activity was needed in neutral HP comparatively to extension and flexion to produce more force and attributed the differences found to differences in muscle length and muscle lever arm length.

Watson and Trott [38] measured the angle C7-tragus-horizontal (used as a surrogate measure for forward HP) in 30 participants with headache and 30 asymptomatic participants and the strength and endurance of the upper cervical flexors. They found lower endurance values of the upper cervical flexors corresponded to lower values of the C7-tragus-horizontal angle. Lower values of this angle indicate a more forward HP. This may suggest that the stabilizing role of the deep flexors is compromised as forward HP increases. Suryanarayana [39] quantified isometric cervical strength at neutral, 25%, 50% and 75% of flexion and extension in 39 volunteers without pain. They found that cervical strength was highest in the neutral posture for both flexion and extension and significantly decreased with an increasing angular deviation of the neck, indicating a mechanical advantage for the cervical musculature when the head and neck are in a neutral alignment. A recent study conducted by Peolsson et al. [40] used ultrasound to evaluate the effect of three different HPs (neutral, flexed and forward HP) on the activity of the dorsal neck muscles when resting and when performing a lifting task with the upper limb in healthy subjects. Results demonstrated significantly greater muscle deformation (indicative of greater muscle activity) induced by flexed and forward HPs, compared to the neutral HP for all dorsal neck muscles at rest (muscles assessed were: upper trapezius, splenius, semispinalis capitis, semispinalis cervicis and multifidus) and a significant interaction between HP condition and activity (rest or lifting). When comparing muscle deformation between rest and lifting, significant changes in muscle deformation were only evident when lifting was performed in the neutral HP. Study authors, suggest that these findings could be explained by high levels of muscle deformation in flexed and forward HP even before the lifting task was initiated. This suggests that HP might modulate loading on the neck structures when upper limb activities are performed. These results are in agreement with those of Elliot [41] and Nimbarte et al. [42] who found that HP influences dorsal neck muscle activity and patterns of recruitment in supine lying with the head loaded and unsupported, during upper limb lifting tasks, respectively.

In addition, HP seems to influence the activity of scapular muscles. Weon [43] assessed the effects of simulated forward HP on the activity of the scapular upward rotators during loaded isometric shoulder flexion in the sagittal plane, in asymptomatic participants. Participants performed shoulder flexion in both simulated forward HP and neutral HP. Significantly increased EMG activity in the upper trapezius and lower trapezius, and significantly decreased EMG activity in the serratus anterior were found during loaded isometric shoulder flexion with forward HP.

If HP deviations from the norm contribute to increase muscle activity while decreasing the ability to produce force and disturbing motor control, this may increase muscle fatigue as well as the internal forces acting on neck sensitive structures leaving individuals vulnerable to injury and pain.

Head Posture and Compressive Loading

The weight of the head and muscle forces create compressive loading that is supported by the cervical spine. [44] It is believed that this force and the consequent intra-disc pressure are affected by different HPs [45] and this is supported by the findings of studies described in the previous section (head posture and muscle activity and force/strength). Bonney and Corlet

[46] investigated the effect of one hour sitting with the head at 0°, 20° and 40° of flexion and found significant differences in the length of the spine between 0° and 20° of flexion (a decrease of 0.88 mm) and between 0° and 40° of flexion (a decrease of 1.65 mm), concluding that flexed HPs load the cervical spine more. This is consistent with the results of the studies described previously reporting that HP deviations from the norm generally increase muscle activity. [36-42] The increase in muscle activity may result in an increase in the compressive forces acting on the neck, possibly explaining the decrease in length of the cervical spine when in flexion. The shrinkage of the spine is usually due to a decrease in intervertebral disc height, which results in an increase in the amount of load supported by the zygapophyseal joints, [47] believed to accelerate inter-vertebral disc and zygapophyseal joints degeneration. [47,48] Zygapophyseal joints have been considered to be the most common source of idiopathic NP [49] although there is little evidence on the consequences of HP deviations on the load supported by intervertebral discs, zygapophyseal joints or ligaments in the neck. Studies using mainly in vitro or biomechanical modeling approaches provide some insight on the effect of head flexion on neck internal loads, showing, in general, an increase in the load supported by the anatomical structures in flexion. Goel and Clausen [50] used a biomechanical model to investigate load sharing among anatomical components of a C5-C6 segment. The authors reported that the inter-spinous and capsular ligaments sustained the most strains in flexion and axial rotation respectively and maximum intradiscal pressure reached the highest values under combined compression and flexion. The quantity of load shared by the disc varied from 14% in compression and extension, 68% in right lateral bending, 75% in left axial rotation, 88% in compression, and 113% in compression and flexion. Snijders et al. [51] used a biomechanical computer model, reporting that during flexion, muscle forces and joint reaction forces increase, except the force between the odontoid and the ligamentum transversum atlantis; the joint reaction forces on the levels C0-C1, C1-C2 and C7-T1 reach minimum values in extension. It has been shown that stretch or load applied to ligaments and discs can change their mechanical properties, affect their function and result in micro-damage in the lumbar spine. [52,53] It can be speculated that changes in tissue load induced by HP deviations may have the same consequences for neck structures contributing to the onset or maintenance of NP.

Head Posture and Degenerative Changes in the Neck

Despite evidence that mechanical changes relate to HP as outlined above, there appears to be no clinical studies exploring whether there is a relationship between HP deviations, an increase in loading and the consequent development of pathology in the neck. However, studies investigating the effect of cervical spine fusion on adjacent-level intradiscal pressure found a marked increase in intradiscal pressure and also an increase in the rates of disc degeneration in segments adjacent to fusion. [48,54] Additionally, evidence from studies investigating the effect of forces on biologic tissues seems to indicate that a variety of changes occur to discs in response to joint loading, including: i) volumetric changes, ii) fluid flow changes, iii) pressure changes, iv) electrokinetic changes and v) deformations in tension, compression or shearing configurations. These factors are considered to be important regulators of cellular metabolism in the disc [55] and therefore, it is likely that overload resulting from daily life activities may adversely affect the nutrition and remodeling changes occurring in the intervertebral discs and consequently contribute to degenerative changes. [55,56] Stokes and Iatridis [57] conducted a review examining the relationship between mechanical loading and spinal disc degeneration. Reviewing evidence from biomechanics, epidemiology, animal models and intervertebral disc physiology, it was concluded that any abnormal loading conditions (including overload and immobilization) could probably produce tissue trauma and/or adaptive changes, resulting in pain. It is unknown whether HP deviations affect the biologic processes occurring in the disc or other anatomical structures. However it can be speculated that if it has the potential to affect the internal forces acting in the neck structures, it is likely to affect their nutrition and remodeling, leading to changes that may facilitate tissue injury and/or prevent complete healing of an existing injury.

Head Posture, Changes in Muscle Blood Flow and the Accumulation of Metabolites

No studies investigating the effect of HP on muscle blood flow or on the concentration of algesic substances were found. However, there is some evidence to support that low level muscle contractions of the trapezius may lead to the accumulation of algesic substances. Boix et al. [58] demonstrated that sustained contractions of the trapezius to maintain shoulder abduction in asymptomatic individuals results in increased bradykinin and pain intensity. Therefore, studies are needed to ascertain whether changes in muscle activity associated with HP can affect muscle blood flow and accumulation of metabolites.

Head Posture and Cervical Range of Motion

There is evidence to suggest that HP deviations from the norm may affect the range of motion available in the cervical spine. Walmsley et al. [59] investigated the effect of five different initial HPs on rotation range of movement in 60 asymptomatic participants, reporting a significant decrease on rotation range of motion when the movement was initiated in flexion, extension, retraction or protraction when compared to neutral. Similar results were obtained by Edmondston et al. [60] who reported right and left rotation and right and left side-flexions to be significantly decreased when the movement was initiated in retraction and forward HP compared to neutral HP. The stretch of the ligaments induced by HP deviations, in addition to anatomy, is likely to limit the movement when compared to the neutral HP. [60] Similar results were found by De-la-Llave-Rincón [61] and Quek et al. [62] The first authors found that the smaller C7-tragus-horizontal angle (reflective of a greater forward HP), the smaller the range of motion in women with carpal tunnel syndrome and 25 matched healthy women. The latter explored the mediating effects of forward HP on the relationship between thoracic kyphosis and cervical mobility in 51 older adults with neck pain with or without referred pain, numbness or paraesthesia. Thoracic kyphosis was measured using a flexicurve and forward HP was assessed via the C7-tragus-horizontal angle. They found that greater thoracic kyphosis was significantly associated with greater forward HP and greater forward HP was associated with less neck flexion and rotation. Interestingly, participants with idiopathic NP were also found to have a decrease in neck range of motion when compared to asymptomatic individuals. [63]

Head Posture and Proprioception

No studies investigating whether HP affects neck proprioception were found. However, investigations into the effect of HP on postural control and the effect of lumbar spine posture on lumbar position sense have been conducted. Kang et al. [64] compared 2 groups of healthy and young computer based workers and found that the group with more pronounced forward HP had their centre of gravity in a relatively anterior location and had lower balance scores than the group with more normal HP. This suggests that forward HP may affect the position of the centre of gravity and contribute to some disturbance in the balance of healthy adults. However, Silva and Johnson [65] when comparing postural control in a simulated forward HP with postural control in normal HP in healthy young individuals found that simulated forward HP did not disrupt postural control. However, authors hypothesised that this might be due to an increase in muscle co-contraction throughout the body in forward HP conditions to avoid falling as participants commented that balance was harder to maintain in these conditions. However, muscle co-contraction was not measured. It was also hypothesised that a longer period in forward HP is required before postural control is affected as Dolan and Green [66] found a significant increase in lumbar joint repositioning error after 5 minutes in a slouched posture, but no change was noted after 3 seconds in the same posture, suggesting that the effects of a non-optimal posture could be time-dependent. This hypothesis needs to be investigated. While initially, the deviation in HP might contribute to impaired proprioception, [67]

Edmondston et al. [68] suggested that subsequently, proprioceptive impairment may make correction of HP deviation problematic, possibly perpetuating pain as individuals adopt postures that place greater load on the neck structures. Studies have also found that individuals with idiopathic neck pain present impairment in neck proprioception when compared to asymptomatic participants. [69,70]

Taken together, the evidence suggests that HP deviations from the norm contribute to increased neck and scapular muscle activity, disturb normal patterns of recruitment, decrease force production either in static postures or during activities of the upper limb, increase joint compressive loading and decrease neck range of motion. These changes suggest that maintaining normal HP might be advantageous and that it is likely that HP deviations from the norm increase the probability of having neck pain or perpetuate existing NP. Interestingly, studies comparing patients with chronic idiopathic NP and asymptomatic participants have also reported that patients show decreased cervical range of motion, [63,71-73] impaired proprioception, [69,70] reduced neck muscle strength, [71,74,75] reduced neck muscle endurance, [76,77] greater neck muscle fatigue, [78] reduced neuromuscular efficiency [79] and altered muscle control [80,81] when compared to asymptomatic participants. This raises the question as to whether HP might be associated with these impairments.

Measurement of Head Posture

Head Posture Assessment in Clinical Practice

Clinicians seem to use HP assessment through observation for patients with neck pain. In particular, in a survey with 278 physiotherapists, 276 reported to assess HP for patients with neck pain. Of the 276 respondents that reported to assess HP for patients with neck pain, 130 (47.1%) reported to 'always' assess HP for these patients and an additional 80 respondents (29.0%) reported to assess HP for more than 67% of their patients. [82] All respondents that assessed HP reported to do it by observation and the use of qualitative descriptors such as: normal HP, mild deviation, moderate deviation and severe deviation. However, a variety of different procedures seem to be used indiscriminately for HP assessment. HP is assessed with the patient standing, sitting, lying down with face up, in the most painful HP, walking, dressing and undressing from lateral, frontal and posterior viewpoints. Instructions given to patients also vary, including asking patients to adopt as natural a posture as possible, keeping a horizontal look or to look straight ahead. [6] The lack of a standardized protocol for HP assessment may account for these discrepancies, leading physiotherapists to do what they perceive as being optimal for the patient based on their beliefs and clinical experience. However, these inconsistencies among physiotherapists clearly threaten the reliability of HP assessment and therefore hinder teamwork between colleagues. For example, Solow and Tallgren [83] found that different instructions can lead to significantly different HPs and Ferrario et al. [84] showed that HP differs between standing and seated positions. In addition, Edmondston et al. [85] found that different sitting postures induce a different pattern of muscle activation and a different posture of the uppermost segments. In this study, having the lumbo-pelvic region in a neutral position promotes greater activation of the thoracic extensor muscles and significantly reduces the levels of cervical extensor muscle activity associated with a decreased forward HP when compared to sitting in a slouched posture.

Furthermore, in clinical practice, HP is assessed by observation without the aid of a measurement instrument. However, visual inspection of the head has been shown to be neither reliable nor valid. [86-88] We asked 10 physiotherapists to assess head posture of 40 individuals using a 4-point scale: 1) normal, 2) slightly deviated, 3) moderately deviated and 4) severely deviated in two separate occasions. [87] Inter-rater agreement was very poor. When matching the rates attributed to the individuals against objective measures of posture we found that individuals with the same objective measurement were attributed quite different rates and individuals with different objective measures were attributed the same rate. Figure 2 shows a graphical presentation of 3 physiotherapist's rates (scale categories) plotted against the objective measurement (the angle between C7, the tragus of the ear and the horizontal) for the assessment of forward HP. It is unlikely that lack of previous experience using the scale influenced the results as 3 of the 10 physiotherapists reported to assess HP in clinical practice using the same scale. These findings compromise the usefulness of HP assessment for patients with neck pain in clinical practice. [87]

Despite the lack of reliability and validity of HP assessment by observation, HP itself seems to be consistent over time. [89] Therefore, it can be used as part of patients' assessment as long as reliable and valid procedures are used. As previously reported, experimental studies have measured distances and angles between anatomical landmarks as surrogate measures for HP. [14,15,24,26,90,91] The aspect of HP most commonly assessed is forward HP and its surrogate measure most commonly used is the angle between C7, the tragus of the ear and the horizontal. [27] This angle can be measured using a goniometer. This is a cheap and easy to use instrument providing a means to use objective measures of HP in clinical practice [14,92]. Alternatively, clinicians can also measure distances between anatomical landmarks and external references, such as a wall, using a metric ruler as previously described. [25] However, the reliability and validity of cheap, simple and easy to use assessments needs further investigation.

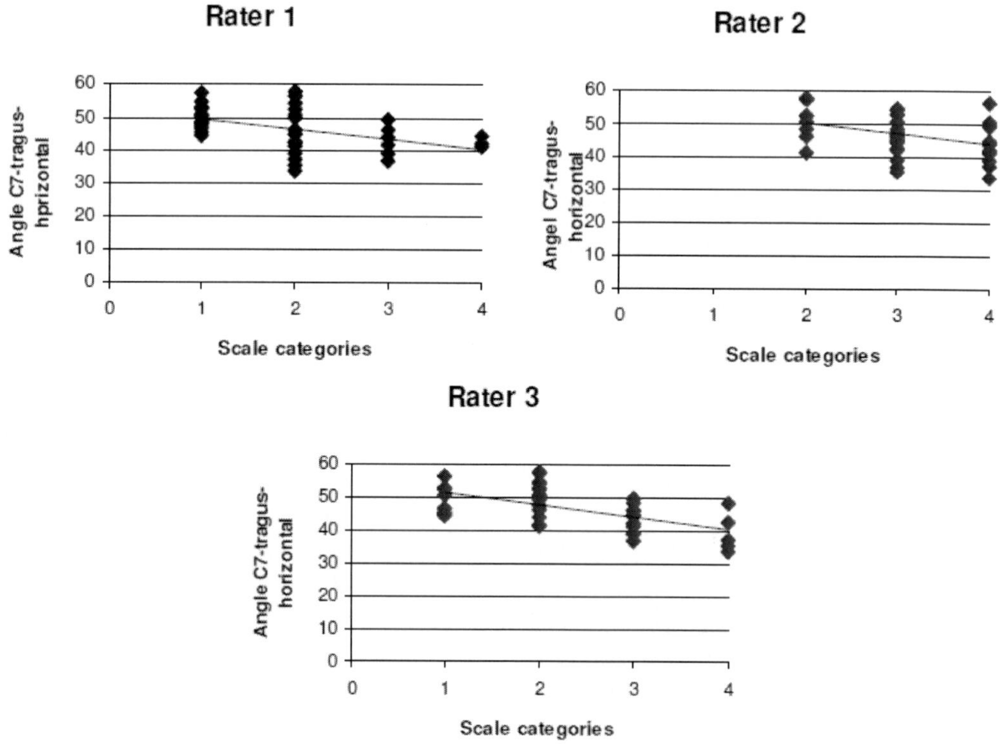

Figure 2. Rates attributed to participants for forward HP plotted against angular values used as a surrogate of forward HP in clinical practice.

The Relationship between Head Posture and Neck Pain

A systematic review from our team aiming to determine whether there were differences in angles or linear distances between anatomical landmarks used as surrogates for HP between individuals with and without neck pain found 13 studies. [19] Of these, 11 studies measured surrogates for forward HP, 6 studies measured surrogates for head extension/flexion and 1 study measured surrogates of head side flexion and rotation. Patients with neck pain were found to have greater forward HP in 4 of 11 studies, greater head extension/flexion in 2 of 6 studies and less side-flexion and less rotation in 1 of 1 study (Table 2). We concluded that there was insufficient good quality evidence to determine whether forward HP, head extension, side-flexion and rotation differ between participants with neck pain and asymptomatic participants. Another study conducted by our team and not included in this systematic review, compared 40 asymptomatic participants with 40 patients with chronic idiopathic neck pain and found a significant difference for forward HP, but not head extension and side flexion, between patients with neck pain and asymptomatic participants. [19] Furthermore, when subgrouping the sample using the sample age mean, a significant difference was found between participants with and without pain aged 50 years or less, but not between those aged more than 50 years (Table 2). These results suggest that age may be relevant to the relationship between HP and neck pain and suggest limitations in the value of measuring HP in older patients presenting with chronic NP. Further research should help clarify this. In addition, the absence of a significant difference between asymptomatic participants and patients with neck pain for head extension and side-flexion suggest that it might not be relevant to assess these aspects of HP for patients with neck pain.

Considering the studies found in our systematic review [27] and our own study [19] there were 6 studies that compared forward HP between patients with neck pain and asymptomatic participants using the angle between C7, the tragus of the ear and the horizontal. [14,19,93–96] Of these 6 studies, 4 found a mean difference between groups [19,93,94,96] that varied between a minimum of 2.2° in seating while playing a computer [94] and a maximum of 6.7° in static standing, [93] consistent with patients with neck pain having a more forward HP than asymptomatic participants. The two studies unable to find a difference used small sample sizes (n=10) at least in one group, raising questions as to whether their results are false negatives. [14,95]

Taken together, these results seem to suggest that measurement procedures, in particular the surrogate measure of forward HP used, might influence study results. Additionally, the assessment of forward HP for patients with NP might be of value as there seems to be differences in the measurements of the angle C7-tragus-horizontal between participants with and without NP. However, the size of the differences are too small, reinforcing the previous suggestion of using objective measures to assess HP in clinical practice as it is highly unlikely that clinicians are able to detect differences of this size either between or within patients.

Head Posture Correction

Is It Possible to Correct Head Posture?

Studies investigating techniques to correct HP have reported statistically significant changes in surrogate measures of forward HP taken before and after exercises regimens in asymptomatic participants, but not for measurements of surrogates of head extension [103,104] (Table 3). Pearson and Walmsley [103] measured the effects of 10, 20, and 30 repeated neck retraction movements on HP in asymptomatic subjects and found a significant change (mean change=4.0°±?mm) in the position and orientation of the tragus of the ear, C2, C5 and T1 consistent with a less forward HP. Harman et al. [104] investigated the effect of a 10-week home exercise program on improving forward HP and extension between a group of asymptomatic participants that received an exercise regimen and a control group that did not receive any intervention. The exercise program consisted of neck extensor and pectoralis major stretches and deep neck flexor and shoulder retractor strengthening exercises. They found a significant mean increase in the angle C7, tragus, horizontal of 2.3° immediately after the exercise program, indicating less forward HP after the treatment. No difference was found for head extension. These studies used healthy individuals so the results might not apply to patients with neck pain and whether the changes reported remain or were reversed after the exercise programme had finished was also not investigated.

Falla et al., [94] Diab and Moustafa [105] and Diab [106] investigated forward HP correction in patients with chronic non severe neck pain, cervical radiculopathy and scoliosis, respectively. The results of all three studies suggest that it is also possible to correct HP in patients. In addition, Diab [106] and Diab and Moustafa [105] showed that corrections are still present at 3 and 6 months, respectively. Falla et al. [94] compared the change in HP measured though the angle C7-tragus-horizontal between subjects with chronic neck pain and control subjects while performing a 10 minute computer task. One group received a 6-week endurance-strength training of the cervical flexor muscles and the other training of the craniocervical flexor muscles. Only the group that received the training of the craniocervical flexor muscles demonstrated an increase in the angle C7-tragus-horizontal (mean=4.4°; 95% CI=3.3°–5.4°), indicative of a decrease in the degree of forward HP.

Diab and Moustafa [105] compared the C7, tragus, horizontal angle in two groups of patients with cervical spondylotic radiculopathy before and after a 10 week intervention. Patients in both the experimental and the control group received superficial heat on the neck for 10 minutes followed by continuous ultrasound on upper trapezius for 10 minutes. The exercise group additionally received a posture corrective exercise programme in the form of strengthening of the deep cervical flexors and shoulder retractors and stretching of cervical extensors and pectoralis. Measurements were taken before starting the treatment, at the end of 10 weeks, and at follow-up period of six months. They found a significant increase in the angle C7, tragus, horizontal in the group that received the exercise regimen, (mean difference after treatment=6.8°; mean difference at 6 months=5.2°), but not in the control group (mean difference after treatment=1.0°; mean difference at 6 months=1.3°).

Diab [106] investigated the effectiveness of forward HP correction in adolescents with idiopathic scoliosis after a 10 weeks regimen of exercises and at three-month follow-up. All the patients received traditional treatment in the form of stretching and strengthening exercises. In addition, patients in the experimental group received a forward HP corrective exercise programme as described previously for Diab and Moustafa [105]. They found a significant difference in the angle C7, tragus, horizontal immediately after treatment (mean difference =7.7°), and at 6 month follow-up (mean difference =7.7°). No significant differences were found for the control group (mean difference after treatment=0.3°; mean difference at 3 months= 0.6°).

Taken together these results suggest that it is possible to correct forward HP and that correction is maintained after the exercise programme has finished. Training of the deep cervical flexors was present in the exercise regimens of all studies that found a significant difference in forward HP after exercise, [94,103–106] suggesting that it might be an essential component of an exercise regimen aiming at correcting forward HP. However, the within subject HP changes induced by the exercise regimens were small.

This is in accordance with the small differences in HP between participants with and without neck pain. Assuming that asymptomatic participants have a normal HP and considering that differences between participants with and without neck pain are small (maximum mean difference in studies that measured the angle between C7, the tragus and the horizontal is 6.7°) it is conceivable to expect similar differences for the within subject forward HP changes after an exercise regimen. Whether it is possible to correct the other components of HP is unknown.

Benefits to the Patient with Neck Pain from Head Posture Correction

In the previous section we discussed evidence that suggest that it is possible to correct forward HP. However, this is relevant to the patients with neck pain only if it results in any benefit, namely in an improvement in signs and symptoms associated with neck pain. Studies comparing pain and nerve root function in patients with neck pain before and after exercises showed conflicting results [94,105,107] (Table 3). Abdulwahab and Sabbahi [107] found that 20 repetitions of head retractions resulted in a significant improvement in the amplitude of the H reflex (a monosynaptic reflex indicative of the reflex arc integrity) and pain intensity of patients with complaints of neck, shoulder and arm pain and hand paraesthesia. Diab and Moustafa [105] compared C6 and C7 amplitude of dermatomal somatosensory-evoked potentials (as an indicative of nerve root function) and pain between two groups of patients with spondylotic radiculopathy after a treatment intervention. Both groups received a similar intervention, except for forward HP corrective exercises (as described in the previous section). They reported a significant mean difference between the exercise and control groups for measured variables immediately after treatment and at 3-month follow-up, consistent with a greater improvement in the group that received the corrective exercises. These results suggest that correcting forward HP in patients with cervical radiculopathy results in short and long term benefits for the patients in terms of pain and root function improvement. In contrast, Falla et al. [94] found a statistically significant improvement in the pain intensity reported in two different groups of patients with chronic neck pain after two different exercise regimens (endurance strength training of the neck flexors and training of the craniocervical flexors). However, only the group that received the training of the craniocervical flexor muscles demonstrated an increase in the angle between C7, the tragus and the horizontal (indicative of a decrease in the degree of forward HP). The contradictory results from Falla et al. [94] and Abdulwahab and Sabbahi [107] and Diab and Moustafa [105] suggest that HP correction may be more important in certain subgroups of patients with neck pain. Further research is needed to ascertain this; for example, comparing traumatic vs non-traumatic neck pain, acute vs chronic neck pain or neck pain vs pain in the neck and arm.

Conclusion

Taken together the evidence reviewed in this chapter support the usefulness of forward HP assessment for patients with neck pain, but not for the other components of HP. This lack of support for head extension/flexion, side-flexion and rotation is not due to evidence of their lack of utility but rather to the lack of studies investigating these components.

The arguments in favor of the clinical usefulness of forward HP assessment for patients with neck pain are diverse:

- There is biomechanical and physiological evidence in favor of normal HP vs. abnormal HP;
- Forward HP is reliable for over time comparisons;
- There are differences between patients with neck pain and asymptomatic participants, consistent with patients having a more forward HP;
- It is possible to correct forward HP;
- Correcting forward HP results in benefits to the patient.

However, the small size of the differences found between patients with neck pain and asymptomatic participants and before and after exercise regimens aiming at correct forward HP, measurement procedures must be objective, valid and reliable. Otherwise, clinicians will not be able to assess forward HP changes within or between patients. Additionally, assessment procedures must follow a standardized protocol as the posture of other segments or the indications given to patients affect the degree of forward HP. To our knowledge there is not a standardized protocol for forward HP assessment in clinical practice. Therefore, and based on the available literature and on studies conducted by our team, we suggest that the following procedures should be used when assessing HP for patients with neck pain as a means to standardize the assessment:

1. The patient should be adequately free from clothes;
2. Standing without shoes in a firm surface;
3. Feet slightly apart (approximately shoulder width);
4. Patients are asked to flex and extend their head, decreasing the amplitude of movement in a progressive manner until they stop in what they feel is their natural/usual HP;
5. When they stop moving their head, measurements with an objective tool, such as a goniometer, are taken for the angle C7, tragus, horizontal;
6. Measurement procedures should be taken 3 times and the average used for posterior comparisons;
7. Additionally and using the same procedures from points 1 to 5, the effect of correcting HP towards what is perceived as being a more optimal HP can also be assessed.

Despite the growing body of evidence on HP and neck pain there is much that needs to be undertaken:

- To investigate whether the usefulness of forward HP assessment differs between subgroups of neck pain;
- To develop and validate a protocol for forward HP assessment in clinical practice with objective, inexpensive and easy to use instruments;
- To develop normative data for forward HP for age and sex;
- To identify cut of points above or below which forward HP correction might be beneficial;
- To test several regimens of exercises for forward HP correction in order to identify the most effective;

– To explore the clinical usefulness of the other components of HP (head flexion/extension, side-flexion and rotation).

References

[1] Schwartz L. A résumé, with comments, of the available literature relating to posture. *Public Health Rep*. 1927;27:1219-48.
[2] Kendall F, McCreary E, Provance P, Rodgers MM, Romani WA. Muscles testing and function, with posture and pain . Philadelphia: Lippincott Williams & Wilkins; 2005.
[3] Petty NJ. Neuromusculoskeletal examination and assessment. A handbook for therapists. Bath: Churchill Livingstone; 2011.
[4] McKenzie R, May S. The cervical and thoracic spine. Mechanical diagnosis and therapy. New Zealand: Spinal Publications; 2006.
[5] American Physical Therapy Association. Guide to physical therapist practice. *Phys Ther.* 2001;81:9-746.
[6] Silva A, Punt D, Johnson M. Using head posture assessment to inform the management of neck pain: a pilot focus group study. *Int J Ther Rehabil.* 2012;19:98-105.
[7] Kesson M, Atkins E. Orthopaedic medicine. A practical approach. Oxford: Butterworth Heinemann; 2001.
[8] Magee DJ. Orthopaedic physical assessment. Philadelphia: Saunders; 2008.
[9] Fejer R, Kyvik KO, Hartvigsen J. The prevalence of neck pain in the world population: a systematic critical review of the literature. *Eur Spine J.* 2006;15:834-48.
[10] Basmajian J V, De Luca CJ. Muscles alive. Their functions revealed by electromyography. Baltimore: Williams and Wilkins; 1985.
[11] Silva AG, Punt TD, Sharples P, Vilas-Boas JP, Johnson MI. Head posture assessment for patients with neck pain: Is it useful? *Int J Ther Rehabil.* 2009;16:43-53.
[12] Silva AG, Sharples P, Punt D, Johnson M. Systematic review of studies comparing head posture between participants with neck pain and asymptomatic participants. Pain Eur. Congr. Lisbon: European federation of chapters of the international association for the study of pain; 2009.
[13] Bryden L, Fitzgerald D. The influence of posture and alteration of function upon the craniocervical and craniofacial regions. In: von Piekartz H, Bryden L, (eds). Craniofacial dysfunction pain. Manual therapy assessment and management. Oxford: Butterworth-Heinemann; 2001. p. 163-87.
[14] Harrison AL, Barry-Greb T, Wojtowicz G. Clinical measurement of head and shoulder posture variables. *JOSPT.* 1996;23:353-61.
[15] Braun BL. Postural differences between asymptomatic men and women and craniofacial pain patients. *Arch Phys Med. Rehabil.* 1991;72:653-6.
[16] McLean L. The effect of postural correction on muscle activation amplitudes recorded from the cervicobrachial region. *J Electromyogr Kinesiol.* 2005;15:527-35.
[17] Woodhull AM, Maltrud K, Mello BL. Alignment of the human body in standing. *Eur J Appl Physiol.* 1985;54:109-15.

[18] Szeto GP, Straker LM, O'Sullivan PB. A comparison of symptomatic and asymptomatic office workers performing monotonous keyboard work-2: neck and shoulder kinematics. *Man Ther*. 2005;10:281-91.
[19] Silva AG, Punt TD, Sharples P, Vilas-Boas JP, Johnson MI. Head posture and neck pain of chronic nontraumatic origin: a comparison between patients and pain-free persons. *Arch Phys Med Rehabil*. 2009;90:669-74.
[20] Binnie C, Cooper R, Mauguiere F, Osselton J, Prior P, Tedman B. Clinical neurophysiology. New York: Elsevier Science Publishers; 2003.
[21] Rokkedal-Lausch T, Lykke M, Hansen MS, Nielsen RO. Normative values for the foot posture index between right and left foot: A descriptive study. Gait Posture;38:843-6.
[22] Redmond AC, Crane YZ, Menz HB. Normative values for the Foot Posture Index. *J Foot Ankle Res*. 2008;1:6.
[23] The Association of Faculties of Medicine of Canada Public Health Educators' Network. Establishing cut-points what is a normal value [Internet]. Primer. 2013 [cited 2013 Oct 28]. Available from: The association of faculties of medicine of Canada public health educators' network.
[24] Dalton M, Coutts A. The effect of age on cervical posture in a normal population. In: Boyling JD, Palastanga N (eds). Grieve's modern manual therapy of the vertebral column. New York: Churchill Livingstone; 1994.
[25] Hanten WP, Lucio RM, Russel JL, Brunt D. Assessment of total head excursion and resting head posture. *Arch Phys Med Rehabil*. 1991;72:877-80.
[26] Raine S, Twomey LT. Head and shoulder posture variations in 160 asymptomatic women and men. *Arch Phys Med Rehabil*. 1997;78:1215-23.
[27] Silva AG, Sharples P, Johnson MI. Studies comparing surrogate measures for head posture in individuals with and without neck pain. *Phys Ther Rev*. 2010;15:12-22.
[28] Taniguchi A, Ogita K, Murata T, Kuzuhara S, Tomimoto H. Painful neck on rotation: diagnostic significance for crowned dens syndrome. *J Neurol*. 2010;257:132-5.
[29] Oatis CA. Kinesiology: the mechanics and pathomechanics of human movement. Philadelphia: Lippincott Williams & Wilkins; 2004.
[30] Watkins J. Structure and function of the musculoskeletal system. Champaign, IL: Human Kinetics; 1999.
[31] Vasavada AN, Li S, Delp SL. Influence of muscle morphometry and moment arms on the moment-generating capacity of human neck muscles. *Spine*. 1998;23:412-22.
[32] Przybyla A, Mohite A, Blease S, Adams MA. How does posture affect lever arms of neck flexor and extensor muscles? *J Biomech*. 2006;39 S49.
[33] Enwemeka CS, Bonet IM, Ingle JA, Prudhithumrong S, Ogbahon FE, Gbenedio NA. Postural correction in persons with neck pain. Part II: Integrated electromyography of the upper trapezius in three simulated neck positions. *JOSPT*. 1986;8:240-2.
[34] McGill SM, Brown S. Creep response of the lumbar spine to prolonged full flexion. *Clin Biomech*. 1992;7:43-6.
[35] Terry GC, Hammon D, France P, Norwood LA. The stabilizing function of passive shoulder restraints. *Am J Sport Med*. 1991;19:26-34.
[36] Kumar S, Narayan Y, Amell T, Ferrari R. Electromyography of superficial cervical muscles with exertion in the sagittal, coronal and oblique planes. *Eur Spine J*. 2002;11:27-37.

[37] Mayoux-Benhamou MA, Revel M. Influence of head position on dorsal neck muscle efficiency. *Electromyogr Clin Neurophysiol.* 1993;33:161-6.

[38] Watson DH, Trott PH. Cervical headache: an investigation of natural head posture and upper cervical flexor muscle performance. *Cephalalgia.* 1993;13:272-84.

[39] Suryanarayana L, Kumar S. Quantification of isometric cervical strength at different ranges of flexion and extension. *Clin Biomech.* 2005;20:138-44.

[40] Peolsson M, Larsson B, Brodin LA, Gerdle B. A pilot study using tissue Velocity ultrasound imaging (TVI) to assess muscle activity pattern in patients with chronic trapezius myalgia. *BMC Musculoskelet Disord.* 2008;9:127.

[41] Elliott JM, O'Leary SP, Cagnie B, Durbridge G, Danneels L, Jull G. Craniocervical orientation affects muscle activation when exercising the cervical extensors in healthy subjects. *Arch Phys Med Rehabil.* 2010;91:1418-22.

[42] Nimbarte A, Aghazadeh F, Ikuma L, Harvey C. Neck disorders among reconstruction workers: Understanding the physical loads on the cervical spine in static lifting tasks. *Ind Health.* 2010;145-53.

[43] Weon J-H, Oh J-S, Cynn H-S, Kim Y-W, Kwon O-Y, Yi C-H. Influence of forward head posture on scapular upward rotators during isometric shoulder flexion. *J Bodyw Mov Ther.* 2010;14:367-74.

[44] Miura T, Panjabi MM, Cripton PA. A method to simulate in vivo cervical spine kinematics using in vitro compressive preload. *Spine* (Phila. Pa. 1976). 2002;27:43-8.

[45] Borestein DG, Wiesel SW, Boden SD. Low back and neck pain. Comprehensive diagnosis and management. Philadelphia: Saunders; 2004.

[46] Bonney RA, Corlett EN. Head posture and loading of the cervical spine. *Appl Ergon.* 2002;33:415-7.

[47] Adams MA, Freeman BJ, Morrison HP, Nelson IW, Dolan P. Mechanical initiation of intervertebral disc degeneration. *Spine.* 2000;25:1625-36.

[48] Pospiech J, Stolke D, Wilke HJ, Claes LE. Intradiscal pressure recordings in the cervical spine. *Neurosurgery.* 1999;44:379-84.

[49] Manchikanti L, Cash K, Pampati V, Wargo B, Malla Y. Cervical epidural injections in chronic discogenic neck pain without disc herniation or radiculitis: preliminary results of a randomized, double-blind, controlled trial. *Pain Physician.* 2010;13:E265-78.

[50] Goel VK, Clausen JD. Prediction of load sharing among spinal components of a C5-C6 motion segment using finite element approach. *Spine.* 1998;23:684-91.

[51] Snijders CJ, van Dijke GAH, Roosch ER. A biomechanical analysis of the cervical spine in static postures. *J Biomech.* 1991;24:783-92.

[52] Solomonow M, Baratta R V, Banks A, Freudenberger C, Zhou BH. Flexion-relaxation response to static lumbar flexion in males and females. *Clin Biomech.* 2003;18:273-9.

[53] Solomonow M, Baratta R V, Zhou BH, Burger E, Zieske A, Gedalia A. Muscular dysfunction elicited by creep of lumbar viscoelastic tissue. *J Electromyogr Kinesiol.* 2003;13:381-96.

[54] Eck JC, Humphreys SC, Lim TH, Jeong ST, Kim JG, Hodges SD, et al. Biomechanical study on the effect of cervical spine fusion on adjacent-level intradiscal pressure and segmental motion. *Spine.* 2002;27:2431-4.

[55] Setton L, Chen J. Mechanobiology of the intervertebral disc and relevance to disc degeneration. *J Bone Jt Surg.* 2006;88:52-7.

[56] Adams MA, Dolan P. Spine biomechanics. *J Biomech.* 2005;38:1972-83.

[57] Stokes IAF, Iatridis JC. Mechanical conditions that accelerate intervertebral disc degeneration: overload versus immobilization. *Spine.* 2004;29:2724-32.
[58] Boix F, Roe C, Rosenborg L, Knardahl S. Kinin peptides in human trapezius muscle during sustained isometric contraction and their relation to pain. *J Appl Physiol.* 2005;98:534-40.
[59] Walmsley RP, Kimber P, Culham E. The effect of initial headposition on active cervical axial rotation range of motion in two age populations. *Spine.* 1996;21:2435-42.
[60] Edmondston SJ, Henne S-E, Loh W, Østvold E. Influence of cranio-cervical posture on three-dimensional motion of the cervical spine. *Man Ther.* 2005;10:44-51.
[61] De-la-Llave-Rincón AI, Fernández-de-las-Peñas C, Palacios-Ceña D, Cleland J a. Increased forward head posture and restricted cervical range of motion in patients with carpal tunnel syndrome. *JOSPT.* 2009;39:658-64.
[62] Quek J, Pua Y-H, Clark R a, Bryant AL. Effects of thoracic kyphosis and forward head posture on cervical range of motion in older adults. *Man Ther.* 2013;18:65-71.
[63] Woodhouse A, Vasseljen O. Altered motor control patterns in whiplash and chronic neck pain. *BMC Musculoskelet Disord.* 2008;20:90-100.
[64] Kang J-H, Park R-Y, Lee S-J, Kim J-Y, Yoon S-R, Jung K-I. The effect of the forward head posture on postural balance in long time computer based worker. *Ann Rehabil Med.* 2012;36:98-104.
[65] Silva AG, Johnson MI. Gait and posture: Does forward head posture affect postural control in human healthy volunteers ? *Gait Posture.* 2013;38:352-3.
[66] Dolan KJ, Green A. Lumbar spine reposition sense: The effect of a "slouched" posture. *Man Ther.* 2006;11:202-7.
[67] Paris S V. Cervical symptoms of forward head posture. *Top Geriatr Rehabil.* 1990;5:11-9.
[68] Edmondston SJ, Chan HY, Ngai GC, Warren ML, Williams JM, Glennon S, et al. Postural neck pain: an investigation of habitual sitting posture, perception of "good" posture and cervicothoracic kinaesthesia. *Man Ther. 2007*;12:363-71.
[69] Kristjansson E, Dall'Alba P, Jull G. A study of five cervicocephalic relocation tests in three different subject groups. *Clin Rehabil. 2003*;17:767-74.
[70] Rix GD, Bagust J. Cervicocephalic kinesthetic sensibility in patients with chronic, nontraumatic cervical spine pain. *Arch Phys Med Rehabil.* 2001;82:911-9.
[71] Chiu TT, Sing KL. Evaluation of cervical range of motion and isometric neck muscle strength: reliability and validity. *Clin Rehabil.* 2002;16:851-8.
[72] Vogt L, Segieth C, Banzer W, Himmelreich H. Movement behaviour in patients with chronic neck pain. *Physiother Res Int.* 2007;12:206-12.
[73] Wallis J a, Webster KE, Levinger P, Taylor NF. What proportion of people with hip and knee osteoarthritis meet physical activity guidelines? A systematic review and meta-analysis. *Osteoarthritis Cartilage.* 2013; 21:1648-59.
[74] Jordan K. Assessment of published reliability studies for cervical range-of-motion measurement tools. *J Manipulative Physiol Ther.* 2000;23:180-95.
[75] Cagnie B, Cools A, De Loose V, Cambier D, Danneels L. Differences in isometric neck muscle strength between healthy controls and women with chronic neck pain: the use of a reliable measurement. *Arch Phys Med Rehabil.* 2007;88:1441-5.

[76] Lee H, Nicholoson LL, Adams RD, Bae SS. Body chart pain location and side specific physical impairment in subclinical neck pain. *J Manipulative Physiol Ther.* 2005;28:479-86.

[77] Harris KD, Heer DM, Roy TC, Santos DM, Whitman JM, Wainner RS. Reliability of a measurement of neck flexor muscle endurance. *Phys Ther.* 2005;85:1349-55.

[78] Falla D, Rainoldi A, Merletti R, Jull G. Myoelectric manifestations of sternocleidomastoid and anterior scalene muscle fatigue in chronic neck pain patients. *Clin Neurophysiol.* 2003;114:488-95.

[79] Falla D, Jull G, Edwards S, Koh K, Rainoldi A. Neuromuscular efficiency of the sternocleidomastoid and anterior scalene muscles in patients with chronic neck pain. *Disabil Rehabil.* 2004;26:712-7.

[80] Falla D, Jull G, Hodges P. Feedforward activity of the cervical flexor muscles during voluntary arm movements is delayed in chronic neck pain. *Exp Brain Res.* 2004;157:43-8.

[81] Johnston V, Jull G, Souvlis T, Jimmieson NL. Neck movement and muscle activity characteristics in female office workers with neck pain. *Spine* (Phila. Pa. 1976). 2008;33:555-63.

[82] Silva AG, Punt D, Johnson M. A postal survey gathering information about physiotherapists' assessment of head posture for patients with chronic idiopathic neck pain. Pain europe. Congr. Lisbon: European federation of chapters of the international association for the study of pain; 2009.

[83] Solow B, Tallgren A. Natural head position in standing subjects. *Acta Odont Scand.* 1971;29:591-607.

[84] Ferrario VF, Sforza C, Miani A, Tartaglia G. Craniofacial morphometry by photographic evaluations. *Am J Orthod Dentofac Orthop.* 1993;103:327-37.

[85] Edmondston SJ, Sharp M, Symes A, Alhabib N, Allison GT. Changes in mechanical load and extensor muscle activity in the cervico-thoracic spine induced by sitting posture modification. *Ergonomics.* 2011;54:179-86.

[86] Cleland J, Childs J, Fritz J, Whitman J. Interrater reliability of the history and physical examination in patients with mechanical neck pain. *Arch Phys Med Rehabil.* 2006;87:1388-95.

[87] Silva AG, Punt TD, Johnson MI. Reliability and validity of head posture assessment by observation and a four-category scale. *Man Ther.* 2010;15:490-5.

[88] Passier LN, Nasciemento MP, Gesch JM, Haines TP. Physiotherapist observation of head and neck alignment. *Physiother Theory Pract.* 2010;26:416-23.

[89] Silva A, Punt TD, Johnson MI. Variability of angular measurements of head posture within a session, within a day, and over a 7-day period in healthy participants. *Physiother Theory Pr.* 2011;27:503-11.

[90] Raine S, Twomey LT. Posture of the head, shoulders and thoracic spine in comfortable sitting. *Aust J Physiother.* 1994;40:25-32.

[91] Shiau YY, Chai HM. Body posture and hand strength of patients with temporomandibular disorders. *Cranio.* 1990;8:244-51.

[92] Teixeira Z, Lã B, Silva A. Head and Shoulder Functional Changes in Flutists. *Med Probl Perform Art.* 2013;28:145-51.

[93] Lau HMC, Chiu TTW, Lam TH. Clinical measurement of craniovertebral angle by electronic head posture instrument: a test of reliability and validity. *Man Ther.* 2009;14:363-8.

[94] Falla D, Jull G, Russell T, Vicenzino B, Hodges P. Effect of neck exercise on sitting posture in patients with chronic neck pain. *Phys Ther.* 2007;87:408-17.

[95] Visscher CM, De Boer W, Lobbezoo F, Habets L. Is there any relationship between head posture and craniomandibular pain? *J Oral Rehabil.* 2002;29:1030-6.

[96] Yip CH, Chiu TT, Poon AT. The relationship between head posture and severity and disability of patients with neck pain. *Man Ther.* 2008;13:148-54.

[97] Hanten WP, Olson SL, Russel JL, Lucio RM, Campbell AH. Total head excursion and resting head posture: normal and patients comparisons. *Arch Phys Med Rehabil.* 2000;81:62-6.

[98] Szeto GP, Straker L, Raine S. A field comparison of neck and shoulder postures in symptomatic and asymptomatic office workers. *Appl Erg.* 2002;33:75-84.

[99] Lee H, Nicholson LL, Adams RD. Cervical range of motion associations with subclinical neck pain. *Spine* (Phila. Pa. 1976). 2003;29:33-40.

[100] Nilsson BM, Soderlund A. Head posture in patients with whiplash-associated disorders and the measurement method's reliability - a comparison to health subjects. *Adv Physiother.* 2005;7:13-9.

[101] Arvidsson I, Hansson G-Å, Mathiassen SE, Skerfving S. Neck postures in air traffic controllers with and without neck/shoulder disorders. *Appl Erg.* 2008;39:255-60.

[102] Straker LM, O'Sullivan PB, Smith AJ, Perry MC, Coleman J. Sitting spinal posture in adolescents differs between genders, but is not clearly related to neck/shoulder pain: an observational study. *Aust J Physiother.* 2008;54:127-33.

[103] Pearson ND, Walmsley RP. Trial into the effects of repeated neck retractions in normal subjects. *Spine.* 1995;20:1245-51.

[104] Harman K, Hubley-Kozey CL, Butler H. Effectiveness of an exercise program to improve forward head posture in normal adults: a randomized, controlled 10-week trial. *J Man Manip Ther.* 2005;13:163-76.

[105] Diab A a, Moustafa IM. The efficacy of forward head correction on nerve root function and pain in cervical spondylotic radiculopathy: a randomized trial. *Clin Rehabil.* 2012;26:351-61.

[106] Diab A. The role of forward head correction in management of adolescent idiopathic scoliotic patients: a randomized controlled trial. *Clin Rehabil.* 2012;26:1123-32.

[107] Abdulwahab SS, Sabbahi M. Neck retractions, cervical root decompression, and radicular pain. *JOSPT.* 2000;30:4-9.

In: Posture: Types, Exercises and Health Effects
Editor: Sarah A. Curran ISBN: 978-1-63117-252-6

Chapter 3

Head and Neck Posture and Upper Spine Morphology in Relation to the Craniofacial Profile and Orofacial Function

Liselotte Sonnesen, * ***DDS, PhD***
Dr. Odont., Specialist in Orthodontics, Head of Section for Oral Surgery, Orthodontics and Oral Radiology, Department of Odontology, Faculty of Health and Medical Sciences, University of Copenhagen, Denmark

Abstract

Associations between posture of the head and neck, morphology of the craniofacial profile, dysfunction of the masticatory muscles and jaws and obstruction of the upper airway have been previously described. More recently, associations between upper spine morphology and the morphology of the craniofacial profile including the cranial base and associations between posture of the head and neck, and the upper spine morphology have been described. These findings indicate that posture of the head and neck and the upper spine morphology is a factor in the development of the craniofacial profile and orofacial function in terms of dysfunction of the masticatory muscles and jaws and obstruction of the upper airways. This knowledge is clinically valuable not only for the dentist and the orthodontist, but also for the physician and physiotherapist.

Keywords: Head, neck, craniofacial profile, dysfunction, upper spine, morphology

* Author email: alson@sund.ku.dk

Introduction

Previously it has been described that posture of the head and neck was related to the morphology of the craniofacial profile, dysfunction of the masticatory muscles and jaws and obstruction of the upper airway. Recent findings have been added concerning associations between upper spine morphology and the morphology of the craniofacial profile including the cranial base. In addition, new findings have shown the link between posture of the head and neck and the upper spine morphology. Accordingly, it is suggested that upper spine morphology and posture of the head and neck are associated with the development of the craniofacial profile and orofacial function.

The main aim of this chapter is to define a standard procedure to record the head and neck posture and to define upper spine morphology. It also aims to discuss the role of upper spine morphology and head and neck posture in the development of the craniofacial profile and orofacial function.

Standardised Upright Posture

The posture of the head and neck can be evaluated from photographs or from cephalometric radiographs and it is important to have a standard procedure that is reliable and can be reproduced. One method that can fulfil this requirement is lateral cephalometric radiographs recorded in the standardised upright posture. [1-5] This method is defined as "a posture of the head and neck that is determined by the subjects' own postural control system". Accordingly, the posture of the head and neck can be defined in two ways: with or without external reference. The "self-balance position" is without external reference (proprioceptive system) and the "mirror position", with external reference (proprioceptive and visual system). This method is reproducible without systematic error and with a method error that is insignificantly small. [6]

To record the lateral cephalometric radiograph in the standardised upright posture, it needs only a few additional requirements from the basic conventional requirement of the cephalometric radiographs. These additional requirements are rehearsal outside the cephalometer and the actual positioning in the cephalometer. The rehearsal outside the cephalometer consists of the positioning of the body, the orthoposition, the positioning of the head, and the self-balance position. The orthoposition, a position of transition from standing to walking, is rehearsed by walking on the spot, raise and lower shoulders, heels together and let the arms hang. The self-balance position is rehearsed by tilting the head up and down with decreasing amplitude (Figure 1).

The actual positioning in the cephalometer consists of positioning of the feet, body and head in the mirror position or self-balance position. The operator places a foot in front or behind the patient's feet. The patient is then instructed to move slightly forwards or backwards to contact the operator's feet. Do not grab or push the patient's head or neck with the hands (Figure 2). Finally, ear rods are gently inserted into the external part of the meatus in order to enable the subject to stabilize the head posture during exposure and a check is made for possible lateral tilt or rotation of the head by means of a light beam cross-projected onto the face (Figure 3).

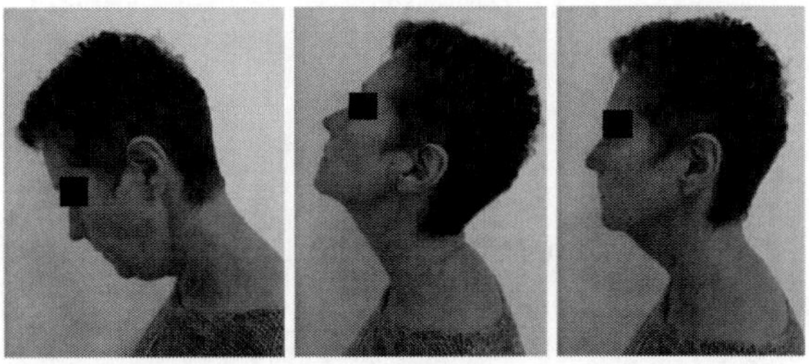

Figure 1. Positioning of the body and head. If necessary the rehearsal outside the cephalometer is repeated by tilting the head up and down with decreasing amplitude.

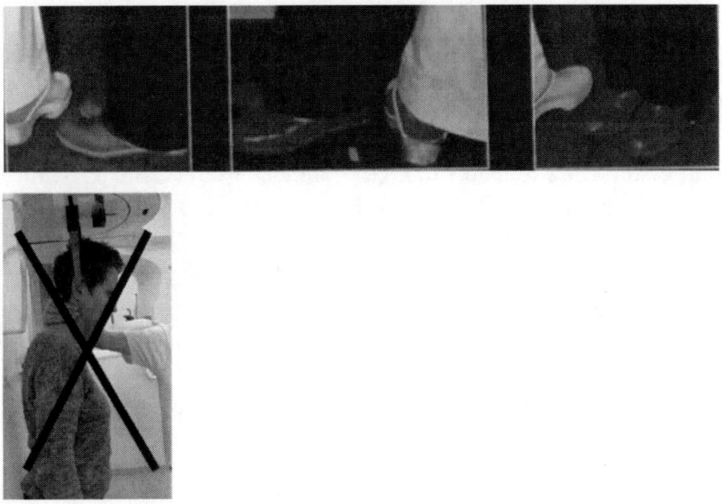

Figure 2. Positioning of the feet in the cephalometer. The operator places a foot in front or behind the patient's feet. The patient is then instructed to move slightly forwards or backwards to contact the operator's feet. Do NOT grab or push the patient's head or neck with the hands.

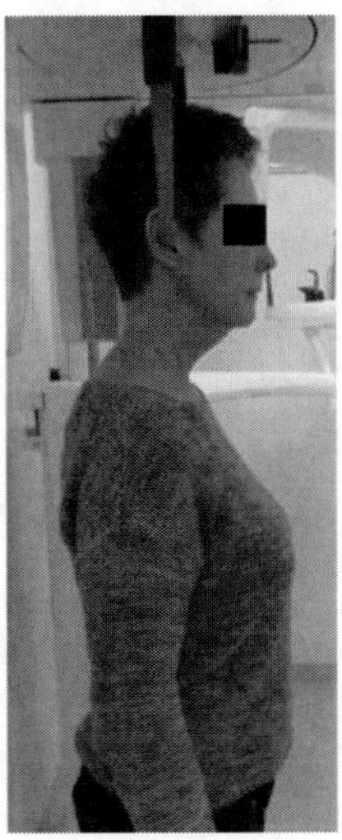

Figure 3. Subject standing in orthoposition in the self-balanced position. Ear rods are gently inserted into the external part of the meatus in order to stabilize the head posture during exposure.

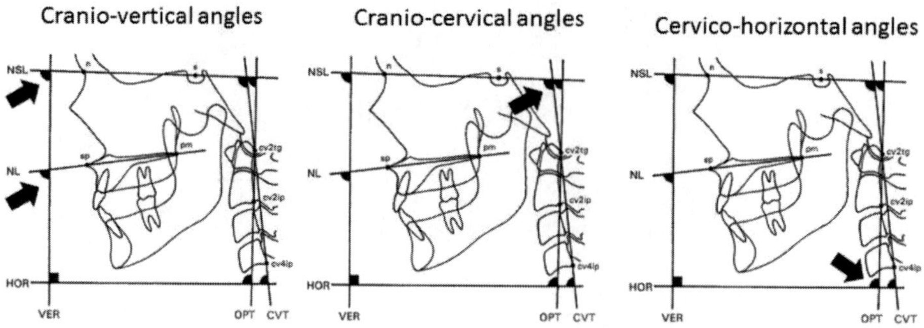

Figure 4. Variables describing the posture of the head and neck.

The variables that describe the posture of the head and neck consist of three main categories [1-3] (Figure 4):

1) Posture of the head related to an environmentally determined vertical or horizontal line, i.e., the cranio-vertical angles (NSL/VER, NL/VER);
2) Posture of the head related to a line representing the upper spine i.e., the cranio-cervical angles (NSL/OPT, NL/OPT, NSL/CVT, NL/CVT);
3) The upper spine inclination expressed in relation to the environmentally determined true horizontal, i.e., the cervico-horizontal angles (OPT/HOR, CVT/HOR).

Terms that are related to these variables:

a) Extension of the head means a raised position of the head in relation to the upper spine or true vertical, i.e., large cranio-cervical angle and cranio-vertical angle respectively;
b) Forward inclination of the upper spine means a small cervico-horizontal angle.

Upper Spine Morphology

The upper spine morphology can be obtained from conventional 2D cephalometric radiographs or from 3D cone beam computed tomography (CBCT). One method to describe the upper spine morphology on either cephalometric radiographs or on CBCT is by visual assessment of the first five cervical vertebral units. The morphological deviations are divided into two categories "Posterior arch deficiency" and "fusion anomalies" [7] (Figure 5):

1) Posterior arch deficiency consisted of *partial cleft*: failure of the posterior part of the neural arch to fuse and *dehiscence*: failure of part of a vertebral unit to develop (Figure 5).
2) Fusion anomalies consisted of *fusion*: fusion of one unit with another at the articulation facets, neural arch or transverse processes, *block fusion*: fusion of more than 2 units at the vertebral bodies, articulation facets, neural arch or transverse

processes and *occipitalization*: assimilation either partial or complete of the atlas (C1) with the occipital bone (Figure 5).

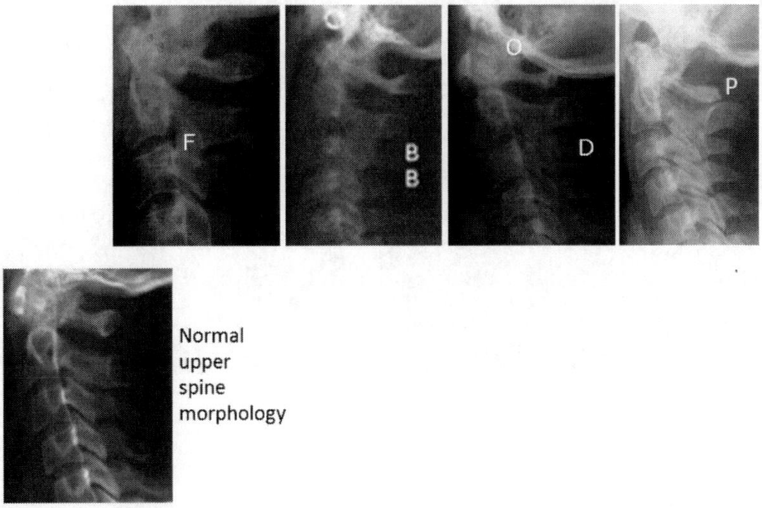

Figure 5. Upper spine morphological deviations and normal upper spine morphology illustrated on lateral cephalometric radiographs. P: partial cleft, D: dehiscence, F: fusion, B: block fusion, O: occipitalization.

The reliability of the visual assessment on either cephalometric radiographs or CBCTs is reproducible without systematic error and with a method error that is insignificantly small. [8, 9]

Posture, Upper Spine Morphology and Craniofacial Profile

Many cross-sectional studies agree on the relationship between posture of the head and neck and craniofacial structures. [3, 10-15] When the head is extended in relation to the upper spine (extended craniocervical posture) an increased anterior facial height, reduced sagittal jaw dimensions, and a steeper inclination of the mandible are generally observed (long craniofacial profile, Figure 6). When the head is flexed in relation to the upper spine there is, on average, a shorter anterior facial height, larger sagittal jaw dimensions and a less steep inclination of the mandible (square craniofacial profile, Figure 6).

It has likewise been demonstrated in a longitudinal study [16] that growth changes in cranio-cervical posture are related to corresponding changes in the growth pattern of the facial skeleton. In other words, a reduction of the cranio-cervical angel is seen in connection with a larger-than-average forward rotation of the mandible and an increase in cranio-cervical angle is seen in connection with a reduced forward rotation of the mandible. Since longitudinal studies do not provide information about cause and effect, a predictive study was carried out to see if it was possible to predict postural changes from mandibular shape, or if it would be possible to predict mandibular growth rotation from cranio-cervical posture. [17]

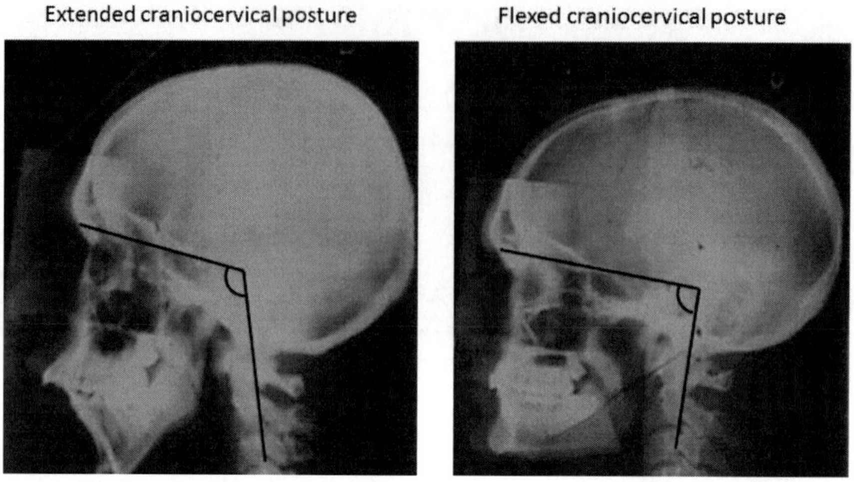

Figure 6. Illustration on lateral cephalometric radiographs of the head and neck posture in relation to the craniofacial profile.

It was found that prediction of the development in the individual child was possible for children with extreme cranio-cervical postural relations. This means that an extreme small cranio-cervical angle (< 79 degrees) was followed by an larger-than-average forward true rotation of the mandible and on average a more horizontal facial growth pattern, where as an extreme large cranio-cervical angle (> 113 degrees) was followed by a less-than-average forward true rotation of the mandible and on average a vertical facial development. These findings do indicate that posture, or factors determining posture, influences the direction of the growth of the face to same extent.

Recently associations between posture of the head and neck and upper spine morphology have been found. [18, 19] The studies showed that the cervical lordosis was significantly more curved in subject with fusion than subjects without fusion (Figure 7). Furthermore, the inclination of the upper spine was more backwards (the cervico-horizontal posture) in subjects with fusion than subjects without fusion (Figure 7) and the head in relation to the upper spine (the craniocervical posture) was more extended in subjects with occipitalization.

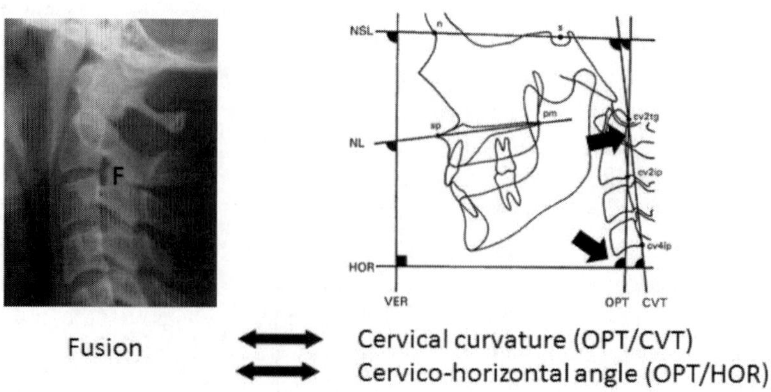

Figure 7. Illustration of associations between head and neck posture and fusion of the upper spine.

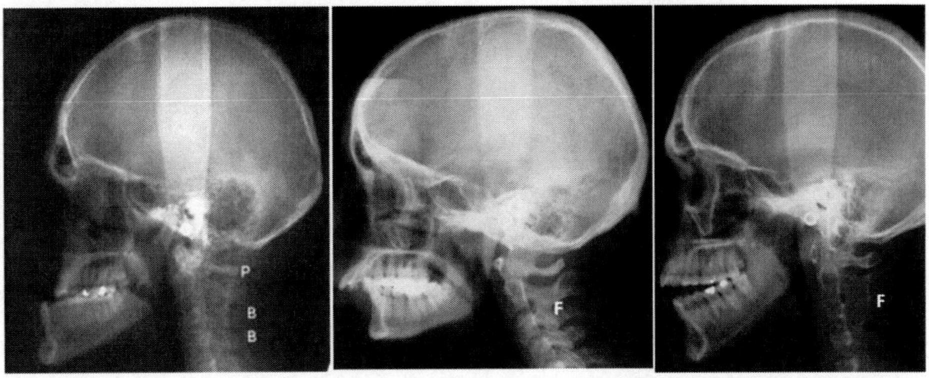

Figure 8. Illustrations on lateral cephalometric radiographs of associations between upper spine morphological deviations severe skeletal malocclsion traits and craniofacial profile. P: Partial cleft, B: Block fusion, F: Fusion between two cervical vertebrae.

Furthermore, a series of studies have determined associations between upper spine morphology and the craniofacial profile in adult patients with severe skeletal malocclusion traits [8, 20-24] (Figure 8). In these studies, it was revealed that fusion of the upper spine was associated with the craniofacial profile. A significant association between fusion and a large cranial base angle, between fusion and retrognathia of the jaws and between fusion and inclination of the jaws was found in patients with severe skeletal malocclusions. This indicates that there might be an association between upper spine morphology, posture of the head and neck and the development of the craniofacial profile including the cranial base.

Two main explanations have been given for the associations between the upper spine, the cranial base and the craniofacial profile: a functional and an embryological approach. The functional approach is the "soft-tissue stretching hypothesis". [25] This hypothesis proposes that a larger cranio-cervical angle (extension of the head) results in a larger distance between the mandible and the sternum, and thus in a stretching of the soft-tissue envelope of the face and neck. The increased tension in the soft-tissue layer will exert slightly higher forces on the facial skeleton. [26] When active over a long period of time during growth, the forces will restrict the forward growth of the maxilla and the mandible, and redirect in a more caudal direction. Such a mechanism would explain the effect of extension of the cranio-cervical posture on the development of the facial skeleton and in particular, of the mandible.

The other approach is the embryological hypothesis regarding the shared origin of the spine and the posterior part of the cranial base to which the jaws are attached. In the early embryogenesis, the notochord develops in the human germ disc and determines the development of the spine, especially the vertebral bodies, and also the basilar part of the occipital bone in the cranial base (the posterior part of the cranial base). [27-31] The para-axial mesoderm forming the vertebral arches and remaining parts of the occipital bone are also formed from notochordal inductions. Therefore, a deviation in the development of the notochord may influence the surrounding bone tissue in the spine as well as in the posterior part of the cranial base. On cephalometric radiographs and CBCTs it can be observed that the bone tissue formed around the notochord are the vertebral bodies and the basilar part of the occipital bone. The shared origin of the spine and posterior part of the cranial base is the basis for the new hypothesis of associations between the spine and the cranial base to which the jaws are attached. [32, 33] From early studies we also know that the angulation of the cranial base influences the growth direction of the jaws. [34, 35] Children with a large cranial base angle develop a larger inclination of the jaws and retrognathia of the jaws (long craniofacial profile). [34, 35] These findings correspond to the craniofacial profile seen in subjects with fusion of the upper spine and an extended posture of the head. [23, 24]

Posture, Upper Spine Morphology and Orofacial Function

The upper spine posture and morphology is not only related to the craniofacial profile, but also to the orofacial function. In this chapter, two aspects of the orofacial function are discussed: Dysfunction of the masticatory system in terms of temporomandibular disorders (TMD) and obstruction of the upper airways. The term of TMD refers to symptoms and signs associated with pain and functional and structural disturbances of the masticatory system, especially the temporomandibular joints (TMJ) and the masticatory muscles. Textbooks and clinicians generally agree that the most important symptoms and signs of TMD are headache, tenderness in the masticatory muscles and the TMJ, reduced or impaired mobility of the mandible, and TMJ sounds. [36] Symptoms and signs of TMD to a certain extent overlap symptoms and signs of upper spine disorders, and clinical observations of a forward head posture in subjects with TMD have already been reported in the 1950s. [37] In more recent studies, a forward head posture (FHP) has been defined as a small value of the angle between a horizontal line and a line from the tragus or the corner of the eye to the spinal process of CV7, assessed clinically or measured on lateral photographs. Studies of FHP have so far reported conflicting results of associations between posture of the head and neck and TMD. [38-42] This lack of consistency of the findings may to some extent have been due to the fact that skin surface measurements of posture do not reflect the actual postural relationship between the bony components of the head and neck. [43] By using a more accurate technique of cephalometric radiographs recorded in the standardized upright posture, it was possible to confirm the Physiotherapists' clinical observations of relationships between TMD and the posture of the head and neck. [5] Subjects with difficulty in jaw opening, clicking and locking of the jaw and asymmetric jaw opening movement showed a marked forward inclination of the upper spine. Furthermore, subjects with locking of the jaw and asymmetric jaw opening movement also had an extended cranio-cervical posture (Figure 9).

Various explanatory models for this relationship have been proposed, but so far no studies have documented whether the symptoms and signs of TMD are the results or the causes of a forward and extended posture of the head and neck, or whether both are triggered by other factors. In subjects with obstruction of the upper airways as in patients with obstructive sleep apnoea (OSA) studies on posture of the head and neck and upper spine morphology have been performed. The prevalence of upper spine morphological deviations ranged between 32% to 46% and the morphological deviations occurred significantly more often in patients with OSA than in subjects with no history of upper airway obstruction and normal craniofacial morphology. [9, 44, 45] Upper spine morphological deviations in OSA patients were described as fusions between two cervical vertebrae, block fusions, occipitalization, partial cleft of the first cervical vertebra (C1) or dehiscence of the third cervical vertebra (C3) and the fourth cervical vertebra (C4) [9, 44, 45] (Figure 10).

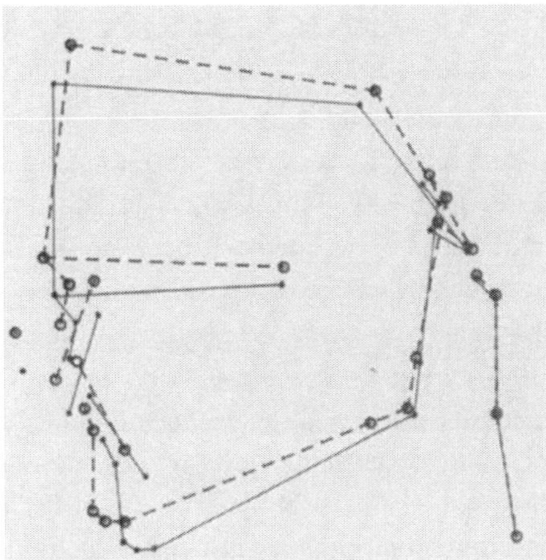

Figure 9. Mean diagrams of lateral cephalometric radiographs of the children with (---) and without (-) locking of the jaw. Superimposition on OPT ($cv2^{ip}$).

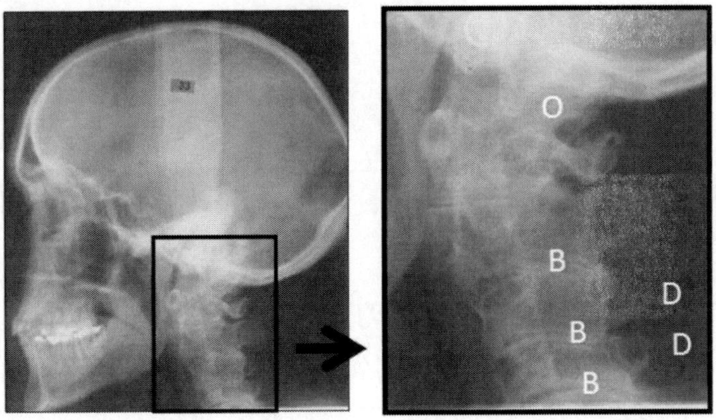

Figure 10. Illustration on lateral cephalometric radiograph of upper spine morphological deviations in patients with obstructive sleep apnoea. O: Occipitalization, B: Block fusion, D: Dehiscence.

In addition, the craniofacial profile of OSA, patients with block fusions in the cervical vertebrae and fusion of two vertebrae differed significantly from the craniofacial profile of other OSA patients. [45] The study suggests different craniofacial phenotypes in patients with sleep apnea based on a new sub-classification of the patients according to the upper spine. This may be a valuable basis for analysis of etiology, for diagnostics and treatment of sleep apnoea patients. Furthermore, the head in relation to the upper spine (cranio-cervical posture) was found to be extremely extended in patients with OSA. [46, 47] The findings were considered to reflect a compensatory physiological postural mechanism that serves to maintain airway adequacy in OSA patients in the awake erect posture. [47]

It would appear that the results indicate that there is an association between the posture of the head and neck, upper spine morphology and orofacial function. So far however, the explanations for the associations are still unknown. Thus, the extended posture of the head in relation to the upper spine in patients with obstructive sleep apnoea is considered to be due to a compensatory postural mechanism that serves to maintain airway adequacy.

Conclusion

Posture of the head and neck and upper spine morphology can be evaluated on cephalometric radiographs and CBCTs recorded in the standardised upright posture. On average, an extended posture of the head in relation to the upper spine as well as fusion between two cervical vertebrae was associated with a long craniofacial profile. On average, TMD were seen in connection with a marked forward inclination of the neck and an extended cranio-cervical posture. On average, an extended posture of the head and a significantly larger occurrence of upper spine morphological deviations were seen in patients with obstruction of the upper airways as in patients with obstructive sleep apnoea. The results indicate that posture of the head and neck and upper spine morphology is a factor in the development of the craniofacial profile and orofacial function in terms of TMD and obstruction of the upper airways. This knowledge is clinically valuable, not only for the dentist and the orthodontist, but also for the physician and physiotherapist.

References

[1] Solow B, Tallgren A Natural head position in standing subjects. *Acta Odontologica Scandinavica.* 1971;29:591-607.

[2] Solow B, Tallgren A. Craniocervical posture in relation to skeletal and dentoalveolar morphology. *Transactions of the European Orthodontic Society.* 1975; p. 231-35.

[3] Solow B, Tallgren A. Head posture and craniofacial morphology. *Am J Physical Anthropology.* 1976;44:417-35.

[4] Solow B, Sonnesen L. Head posture and malocclusion. *Eur J Orthod dontics.* 1998;20: 685-93.

[5] Sonnesen L, Bakke M, Solow B. Temporomandibular disorders in relation to craniofacial dimensions, head posture and bite force in children selected for orthodontic treatment. *Eur J Orthod.* 2001;23:179-92.

[6] Siersbaek-Nielsen S, Solow B. Intra- and interexaminer variability in head posture recorded by dental auxiliaries. *Am J Orthod.* 1982; 82:50-7.

[7] Sandham A. Cervical vertebral anomalies in cleft lip and palate. *Cleft Palate J.* 1986; 23:206-14.

[8] Sonnesen L, Kjær I. Cervical body fusions in patients with skeletal deep bite. *Eur J Orthod.* 2007; 29:464-70.

[9] Sonnesen L, Jensen KE, Petersson AR, Petri N, Berg S, Svanholt P. Cervical vertebral column morphology in patients with obstructive sleep apnoea assessed using lateral cephalograms and cone beam CT. A comparative study. *Dentomaxillofac Radiol.* 2013; 42:20130060.

[10] Opdebeek H, Bell WH, Eisenfeld J, Mishelevich D. Comparative study between the SFS and LFS rotation as a possible morphogenetic mechanism. *Am J Orthodontics.* 1978;74:509-21.

[11] Marcotte MR. Head posture and dentofacial proportions. *Angle Orthodontics.* 1981; 51:208-13.

[12] Von Treuenfels H. Die Relation der Atlasposition bei prognather und progener Kieferanomalie. *Fortschr Kieferorthop.* 1981;42:482-91.

[13] Solow B, Siersbæk-Nielsen S, Greve E. Airway adequacy, head posture, and craniofacial morhpology. *Am J Orthodontics.* 1984;86:214-23.

[14] Hellsing E, McWilliam J, Reigo T, Spangfort E. The relation between craniofacial morphology, head posture and spinal curvature in 8, 11 and 14-year-old children. *Eur J Orthod.* 1987; 9:254-64.

[15] Huggare J. Head posture and craniofacial morphology in adults from Northern Finland. *Proc Finn Dent Soc.*1989; 82:199-208.

[16] Solow B, Siersbæk-Nielsen S. Growth changes in head posture related to craniofacial development. *Am J Orthodontics.* 1986;89:132-40.

[17] Solow B, Siersbæk-Nielsen S. Cervical and craniofacial posture as predictors of craniofacial growth. *Am J Orthodontics Dentofacial Orthopedics.* 1992;101:449-58.

[18] Sonnesen L, Pedersen CE, Kjær I. Cervical column morphology related to head posture, cranial base angle, and condylar malformation. *Eur J Orthod.* 2007;29:398-403.

[19] Arntsen T, Sonnesen L. Cervical vertebral column morphology related to craniofacial morphology and head posture in preorthodontic children with Class II malocclusion and horizontal maxillary overjet. *Am J Orthod Dentofacial Orthop.*2011; 140:e1-7.

[20] Sonnesen L, Kjær I. Cervical column morphology in patients with skeletal Class III malocclusion and mandibular overjet. *Am J Orthod. Dentofacial Orthop.* 2007;132: 427.e7-12.

[21] Sonnesen L, Kjær I. Anomalies of the cervical vertebrae in patients with skeletal Class II malocclusions and horizontal maxillary overjet. *Am J Orthod Dentofac Orthop.* 2008; 133: 198.e15-20.

[22] Sonnesen L, Kjær I. Cervical column morphology in patients with skeletal open bite. *Orthod. Craniofac. Res.* 2008; 11:17-23.

[23] Sonnesen L. Association between the Cervical Vertebral Column and Craniofacial Morphology. *Int J Dent.* 2010;295728.

[24] Sonnesen L. Cervical vertebral column morphology associated with head posture and craniofacial morphology. *Semin Orthod.* 2012; 18:118-25.

[25] Solow B, Kreiborg S. Soft-tissue stretching: A possible control factor in craniofacial morphogenesis. *Scandinavian J Dental Res.* 1977;85:505-7.

[26] Hellsing E, L´Estrange P. Changes in lip pressure following extension and flexion of the head and a changed mode of breathing. *Am J Orthod Dentofac Orthop.* 1987;91: 286-94.

[27] Müller F, O'Rahilly R. The early development of the nervous system in staged insectivore and primate embryos. *J Comp Neurol.* 1980; 193:741-51.

[28] Kjær I. Human prenatal craniofacial development related to brain development under normal and pathologic conditions. *Acta Odontol Scand.* 1995; 53:135-43.

[29] Kjær I. Neuro-osteology. *Crit Rev Oral Biol Med.* 1998;9:224-44.

[30] Kjær I, Hansen BF. The adenohypophysis and the cranial base in early human development. *J Craniofac Gen Dev Biol.* 1995;15:157-61.

[31] Sadler TW. Embryology of the neural tube development. *Am J Med Gen.* 2005;135C: 2-8.

[32] Sonnesen L, Nolting, Kjær KW, Kjær I. Associations between the development of the body axis and the craniofacial skeleton studied by immunohistochemical analyses using Collagen II, Pax 9, Pax 1, and Noggin antibodies. *Spine.* 2008; 33:1622-26.

[33] Sonnesen L, Nolting D, Engel U, Kjær I Cervical vertebrae, cranial base, and mandibular retrognathia in human triploid fetuses. *Am J Med Genet. Part A.* 2009; 149A:177-87.

[34] Björk A. The face in profile. Sven Tandlak Tidskr. 1947;40:1.

[35] Björk A. Kæbernes relation til det øvrige cranium. In: Lundstrøm A (ed). Nordisk Lärobok I Ortodonti. Stockholm: Sveriges Tandläkarforbunds Förlagsförening. 1975; 69-110.

[36] Bush FM, Dolwick MF (eds). Signs and symptoms. In: The temporo-mandibular joint and related orofacial disorders. J B Lippincott Company, Philadelphia. 1995. p. 35-36.

[37] Perry HT Jr. Facial, cranial and cervical pain associated with dysfunction of the occlusion and articulations of the teeth. *Angle Orthodontist.* 1956; 26:121-28.

[38] Kritsineli M, Shim YS. Malocclusion, body posture, and temporomandibular disorders in children with primary and mixed dentition. *J Clinical Pediatric Dentistry.* 1992;16: 86-93.

[39] Hackney J, Bade D, Clawson A. Relationship between forward head posture and diagnosed internal derangement of the temporomandibular joint. *J Orofacial Pain.* 1993;7:386-90.

[40] Watson DH, Trott PH. Cervical headache: an investigation of natural head posture and upper cervical flexor muscle performance. *Cephalalgia.* 1993;13:272-84.

[41] Lee WY, Okeson JP, Lindroth J. The relationship between forward head posture and temporomandibular disorders. *J Orofacial Pain.* 1995; 9: 161-67.

[42] Visscher CM, de Boer W, Lobbezoo F, Habets LLMH, Naeije M. Is there a relationship between head posture and craniomandibular pain? *J Oral Rehab.* 2002; 29:1030-36.

[43] Johnson GM. The correlation between surface measurement of head and neck posture and the anatomic position of the upper cervical vertebrae. *Spine.* 1998; 23:921-27.

[44] Sonnesen L, Petri N, Kjær I, Svanholt P. Cervical column morphology related to head posture, cranial base angle, and condylar malformation. *Eur J Orthod.* 2007; 29:398-403.

[45] Svanholt P, Petri N, Wildschiødtz G, Sonnesen L, Kjær I. Associations between craniofacial morphology, head posture, and cervical vertebral body fusions in men with sleep apnoea. *Am J Orthod Dentofac Orthop.* 2009; 135:702. e1-9.

[46] Solow B, Ovesen J, Nielsen PW, Wildschiødtz G, Tallgren A. Head posture in obstructive sleep apnoea. *Eur J Orthod.* 1993; 15:107-14.

[47] Solow B, Skov S, Ovesen J, Norup PW, Wildschiødtz G. Airway dimensions and head posture in obstructive sleep apnoea. *Eur J Orthod.* 1996;18:571-9.

In: Posture: Types, Exercises and Health Effects
Editor: Sarah A. Curran
ISBN: 978-1-63117-252-6

Chapter 4

Emotion: The Missing Link in Posture

José Luís Pimentel do Rosário, [*] ***PT, PhD***
Federal University of São Paulo, São Paulo, Brazil

Abstract

Posture is defined as the way we stand and how we ambulate from one point to the next. It is associated with many aspects of human life including our political, emotional and ethical positions. Despite this importance, the study of human posture is relatively new compared to other areas of medical science. Certain deviations in posture can be unsightly and can adversely affect muscular efficiency, as well as predisposing individuals to musculoskeletal pathology. Posture can also alter or be altered by certain psychological conditions.

The following factors can influence posture: biomechanics and the relation of the body with gravity; the vestibular system and balance; the visual system; proprioception, among others. The management of just one of these components in isolation could be perceived as a clinical mistake, since there are many variables that can be present each of which can include the role emotions. Pain or pleasure are described as opposite forms of emotions which in turn can cause changes in action and decisions. The influence of pain in particular has been linked to changes in emotion, and recent studies have confirmed the connection between posture and the human emotional system. Awareness and intervention to address the influence of emotions could assist with postural problems that remain resistant to traditional postural management. The aim of this chapter is to review the aspects of biomechanics and neurophysiology of posture and explore how they can be related to emotions experienced by individuals.

Keywords: Emotion, posture, pain, postural problems, posterior chain

[*] Contact email: joerospt@gmail.com

Introduction

The adoption of upright posture occurred relatively early in human evolutionary history. Three million years ago, the pelvis was transformed from being adapted primarily to arboreal quadrupedality as seen in living apes, to being tailored to terrestrial bipedality as seen in modern humans, with the whole body supported by the feet. [1] An ancient riddle, dating back over 2,500 years, describes postural changes in man: "what is it that walks on four feet in the morning, two at noon, and three at night?" Those who could not answer this riddle to the sphinx were put to death. Greek mythology provided the following answer: "Man, who crawls on four limbs as a baby, walks upright on two as an adult, and walks with the aid of a stick in old age." [2]

This riddle demonstrates how posture is influenced by many aspects of human life. Our posture is characterized by how we behave in life, how we relate to other people, the way we stand and how we move from one point to the next. Our political, emotional or ethical positions are evident in our posture and stance, so too are our viewpoint, standpoint, position, stance, condition, state, mood, and attitude in various situations. In spite of this importance, the study of human posture is relatively new compared to other areas of medical science. Certain deviations in posture can be unsightly and can adversely affect muscular efficiency, as well as predisposing individuals to musculoskeletal pathology. It is also acknowledged that posture can be influenced by some psychological conditions such as sadness and depression. [3-7] For these reasons, posture has been studied by various professionals such as medical practitioners, psychologists and physical therapists, architects, podiatrists, sports scientists as well as others.

In the nineteenth century, based on his observations, Charles Darwin wrote about the role of posture. [8] This study however, only began to emerge in 1927, when Wilhelm Reich, an Austrian physician, paid attention to the fact that body language could be more revealing than the words a patient used. He further acknowledged that body posture was closely related to muscle contractions, which he referred to as 'body armor' that in turn was provoked by emotions. [9] By 1930, the Australian author Frederick Alexander, after curing his own hoarseness and correcting his cervical spine, went on to defend the idea that a better posture would prevent the onset of diseases and body aches. His studies on the alignment of the cervical and lumbar spine to facilitate an organic general improvement were called "The Alexander technique", [10] and is used to the present day by various clinicians (e.g., osteopaths, physical therapists).

In 1947, the kinesiotherapist Françoise Mézières stated that human muscles are completely inter-related and demonstrated that we do not have a single muscle that causes poor posture, but rather chains of muscles (Figure 1) that can create pathology in a specific location. This can be from a generalized tension, whereby the root of the problem can be distant from where the patient feels pain. [11] In the next decade, Ida Rolf began to teach her recently developed technique based on the fascia instead of muscles. [12] It is interesting to note that the patterns of the fascia treated by Ida Rolf and further explained by her students follow a very similar pattern of the Mézières' chains. [13]

The first edition of *"Muscles: Testing and Function"* was released in 1967. New chapters, specifically about posture, were added to later editions and the words 'with Posture and Pain' were also added to the title. The core textbook for many involved in musculoskeletal assessment and rehabilitation and is now in its 5th edition. The prime feature of this textbook focused on the development of an established standard assessment of posture (Figure 2) and drew attention to the fact that only correct posture relieves certain spine, feet and leg pain. [14] In the seventies, Mosche Feldenkrais, an Israeli physicist and martial arts master, explained that body posture can only be improved when there is conscience of the movement. According to Feldenkrais, the improvement of kinesthetic and proprioceptive self-awareness of functional movement and body position could lead to increased function, reduced pain, and greater ease and pleasure of movement. [15]

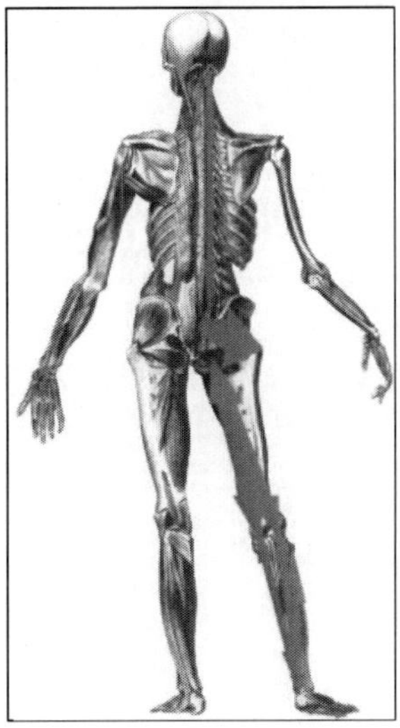

Figure 1. The *posterior chain* according to the *Muscular Chain Therapy* (based on the Mézières method).

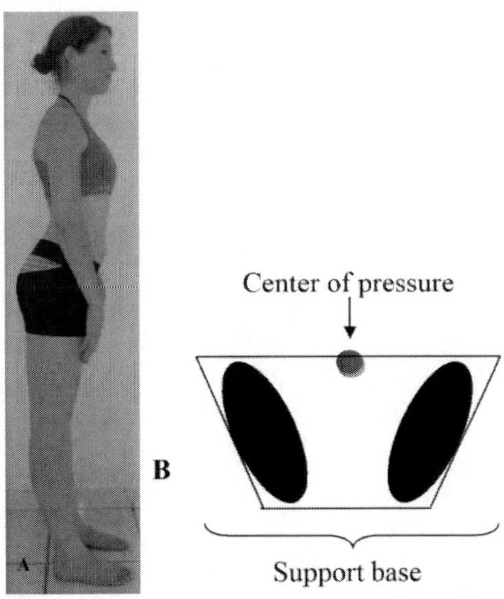

Figure 2. A bad posture alters the relation between the body segments, which may alter the joint mechanics and produce pain (A). This specific alteration is pushing the body forward, which can also move the center of pressure of the body in the same direction (B).

During this period and thereafter, many schools and authors showed an interest in researching human posture. [3,5,6,16-29] In the beginning however, only Reich [9] associated posture and emotions. The aim of this chapter is to review the aspects of biomechanics and neurophysiology of posture and explore how they can be related to emotions experienced by individuals.

Biomechanics

Ideal postural balance exists when there is a perfect distribution of body mass around the center of gravity, with the compressive forces on spinal discs balanced by ligamentous tension. [30] This position creates the least amount of joint stress, requires the least amount of muscular activity and, therefore, is the position of maximum effectiveness. A deviation from this optimum position should be compensated by changes in joint position which, in turn, must be maintained by an increase in muscle activity. Therefore, postural imbalance results in excessive energy consumption. [31]

Ideal posture should not deviate from a plumb line passing sagittally in an anterior or posterior view. In the lateral view, the plumb line should pass slightly anterior to the lateral malleolus, slightly anterior to the axis of the knee joint, and slightly posterior to the axis of the hip joint, the bodies of lumbar vertebrae, the shoulder joint, the bodies of most cervical vertebrae as well as the external auditory canal and slightly posterior to the apex of the coronal suture. [14]

Present day orthopaedic assessment uses the assumption that referred pain in a joint may have its cause in adjacent joints. [32] In terms of posture, this concept can be taken beyond adjacent segments. The misalignment of a body segment will influence the next, which will have a further influence, and is described as a chain reaction. This can occur in the following circumstances:

- When a muscle is in a shortened position to distribute the tension generated in a joint to other muscles;
- When tension is distributed in the fascia and connective tissue; when the position of the center of gravity is altered and compensated at an opposite change of equal magnitude in one or more segments above or below the unaligned segment.

Thus, a deformity that occurs anywhere in the body can be compensated many times throughout the body segments, and the last segment, which cannot be compensated in any other point, is usually the location of the pain. [33]

Spine

Much of the muscle tension generated will be transmitted to the spine, which is a central region with great capacity for compensation at all levels. This tension can arise in the lower limbs (ascendant), or in the head (descendant). [34] It is considered a very important region once it holds the spinal cord. Postural misalignments on the spine can generate pain all over the body and even visceral problems. [35] All postural methods highlight the importance of the spine for the treatment of postural deviations and overall health. Due to its many articulations the spine can show several different alterations in posture, which can be small. These include inclinations or rotations in one or more vertebras, affecting many spinal segments such as hyperlordosis or scoliosis.

Diaphragm and Inspiratory Muscles

The respiratory pump consists of breathing muscles which affect the ribs, scapula, clavicle, sternum and thoracic spine. The fact that breathing can be altered by changes in position, emotional state, activity level, level of alert or sleep stage, illness, posture, muscle strength, airflow resistance, compliance of the lungs and chest wall and even tight clothing means that there are numerous varieties in breathing patterns. [14,36] When posture is altered, the body tends to maintain the chest in an inspiratory position rather than an expiratory position, for survival purposes. [37]

In fact, the shortening of the diaphragm produces postural changes such as elevated ribs and hyperlordosis at L2 level and, with the help of other inspiratory muscles, could cause the elevation of the entire rib cage. [38]

Lower Limbs

When one limb is significantly shorter than the other, the result is referred to as a functional scoliosis. [39] Several idiopathic scoliosis types are caused by a greater torsion in one of the lower limbs, causing the same effect as a smaller limb. This rotation of the lower limb, in turn causes rotation of the hips. When these alterations reach the spine they may produce compensations in all planes, vertebral inclination, rotation and even listesis. Inclinations and rotations are the root cause of scoliosis. [40] Any disturbance at the foundational level of the feet (feet are the interface between the body and supporting surface) can, depending on its features, affect the knee (valgus or varus misalignment or asymmetric rotations) provoking in many cases, spinal alterations such as hyperlordosis, straightening of the spine or scoliosis. [41]

Upper Limbs

Compared to the lower limbs, the upper limbs do not have a marked an effect on posture, but a bundle of tonic muscle connects the forearm to the arm, the shoulder girdle and the neck. All the tension of the upper limbs can be transmitted to the body through the neck, while the opposite can also happen. [37,38]

Temporomandibular Joint

The position of the cervical spine has a close relationship with the temporomandibular joint (TMJ). [42] When the atlanto-occipital joint moves toward flexion, an associated sliding backwards movement occurs in the same joint. When the movement is characterized by extension, it is associated with a sliding forward movement. [43] Forces generated by anterior translation of the atlas with the occipital (extension) lead to a retropulsion of the mandible. In the opposite case, cervical flexion provokes a translation and consequently a jaw antepulsion. [44]

Neurophysiology

Posture is peripherally produced by alpha motor neurons and commonly involves the antigravity muscles. Virtually all systems involved in the regulation of posture and the production of movement exert a direct (monosynaptic) or indirect (via segmental interneurons) influence on the somatic efferent neurons. [45]

Three sensory systems provide information about the body position in relation to gravity and the environment. These are known as the vestibular, proprioceptive and visual systems. The vestibular system provides information about head position in relation to gravity and linear and rotational head movements. Proprioceptors, particularly those associated with the axial joints and muscles (muscle spindles and Golgi tendon organs), provide information about the movement of individual body segments. The visual system provides information about the body position in relation to the external environment. [41,45]

The muscle spindles monitor changes in muscle length and the speed of these changes and play a vital role in sensing position and movement. The spindle may be the cause of a muscle spasm, which is sometimes strong and painful, in the case of cramp, and is sometimes caused by the central nervous system, known as spasticity. [46] These spasms can affect posture even when light and permanent. [47] The Golgi tendon organs monitor intramuscular tension and inhibit muscle contraction with an exaggerated protective function. They are also responsible for muscular relaxation after stretching. [48]

The vestibular system maintains body balance by regulating the antigravity muscles of the neck, trunk and limbs and preventing the body's center of mass from moving away from the support base. It aids by coordinating movements of the head and body, and maintaining the eyes fixed while the head is moving. Balance abnormalities caused by an injury to the labyrinth or vestibular nerve (peripheral vestibular syndrome) usually broaden the support base and can cause the head and trunk to tilt to the side of the injury. [49]

The visual system may be the cause of serious deviations of the neck and trunk. A strabismus, caused by hypofunction of the right lateral rectus muscle in the right eye, can cause a head rotation to the left. This pure example however is rare. Rotation and side bending caused by more than one muscle usually compensate for the problem. Bilateral palpebral ptosis (eyelid closure) requires the carrier to maintain the head in extension so that they can see through the remaining opening. [50]

Body Image

Another important factor is body image, which is defined as a visual and mental image, including feelings of one's own body. Body schema is slightly different and refers to a type of body posture, including the relationship of body parts with each other, and the relationship of the body with the environment. This phenomenon derives from the integration of the tactile, proprioceptive and interoceptive functions. [51,52] Asomatognosia is a disorder which causes a lack of awareness of body structure and the relationship between body parts in the individual. Unilateral spatial neglect is the failure to register and integrate stimuli and perceptions on one side of the body. [53] Common disorders such as anorexia nervosa (characterized by significant weight loss, distorted body image and the pursuit of thinness) are linked to body image. [54-56] Therefore, even small disturbances may affect posture in some way.

Sacks [57] reported a case of a 93-year-old man with Parkinson's disease who exhibited a significant trunk tilt to one side that was almost on the verge of collapse, without realizing it. The author commented that the proprioceptive and visual systems of this patient had affected the individual body image and labyrinthine.

Emotions

Emotions are brief psychophysiological phenomena that have an adaptive aspect in relation to a changing environment. They alter attention, change certain behaviors and activate relevant associative networks in memory. Emotions directly influence perception through the senses, affecting information processing and subsequent actions. [58] Human behavior is strongly determined by emotions, [59] and it is possible to consider them as the primary motivational system of human beings. [60]

According to Aristotle, [61] emotions are reactions accompanied by pain or pleasure that lead to alterations in people, causing them to change their mind. An emotion can also be defined as the integration of particular sets of neurochemical, motor and mental processes, with a number of structural characteristics that distinguishes it from other emotions. [60,62] There is evidence [59,63] of distinct activity patterns in the Autonomic Nervous System (ANS) for anger, fear and sadness. [64] An emotion can quickly organize the physiological responses of different biological systems such as facial expressions, muscle tone, the voice, ANS activity and endocrine activity.

These alterations are intended to produce the best conditions for an effective physical response. [60,64] Siegel [65] stated that moods change the physical state by means of the central nervous system, the endocrine system and the immune system. According to the same author, every tissue and organ is influenced by a complex interaction between hormones that obey the hypothalamus and the pituitary. The hypothalamus regulates the majority of the body's unconscious life processes, such as heartbeat, respiration and blood pressure, among others. Nerve fibers enter the hypothalamus from almost all regions of the brain. Thus, the emotional and intellectual processes which occur in the brain affect the body. This may be one way that emotions influence the physical state and vice versa. Emotions, as well as posture and facial expressions are usually momentary. [66] However, it is notable that people often repeat certain emotional and postural patterns.

Darwin postulated that changes in habits, such as changing the position of a vase, would have an inherent effect on plants, changing the conformation of the stem of a flower in search of sun that focuses on the new position. [67] The same author argues that this effect would be even more striking on animals and human beings. In terms of the muscles, the law of use and disuse has been scientifically proven in studies of immobilization. [68-70] A very important discussion regarding habit is associated with this law: "How many habitual actions are performed unconsciously, indeed often in direct opposition to our conscious will! Once a habit is acquired, it will often remain constant throughout life". [67] In contrast, what differentiates a habit from an instinct is that the former is acquired throughout life while the latter is usually received at birth and is common to the species. [67].

There is a similarity between humans of all races in tastes, dispositions and habits, as pointed out by Darwin: "This is shown by the pleasure which they all take in dancing, rude music, acting… …in their mutual comprehension of gesture-language—and, as I shall be able to show in a future essay, by the same expression in their features, and by the same inarticulate cries, when they are excited by various emotions". [71] Therefore, some of these gestures or postures can be classified as instincts since they are considered innate to the human species. They are adopted during emotional states and can become habits, which may lead to a mental or physical pathology related to the law of use and disuse.

Darwin divided the instincts into two subclasses: self-preservation instincts (hunger, thirst, lust, revenge etc.) which seek to ensure survival, and social instincts (need for approval and fear of censorship) that lead to obtaining pleasure in society with their peers. These two instincts are constantly battling each other and the stronger instinct directs the actions in a given situation. [71] In seeking such approval or censorship, individuals attain moral concepts, but also psychological marks and traumas.

A number of authors have commented on the human instincts of emotional defense and how they manifest themselves as tension in the muscles. Desires generated by the instincts serve to keep us alive and are accepted in our society. The way these desires affect the individual leads to different actions in different situations. When, for whatever reason, one of these instincts becomes so strong and is transformed into a habit, the range of emotional choices are limited. Likewise, the stress distribution generated by constant conflict also produces a constant influence on the muscles that become accustomed to this state and its new position, leading to a matching posture throughout this process. [72,73]

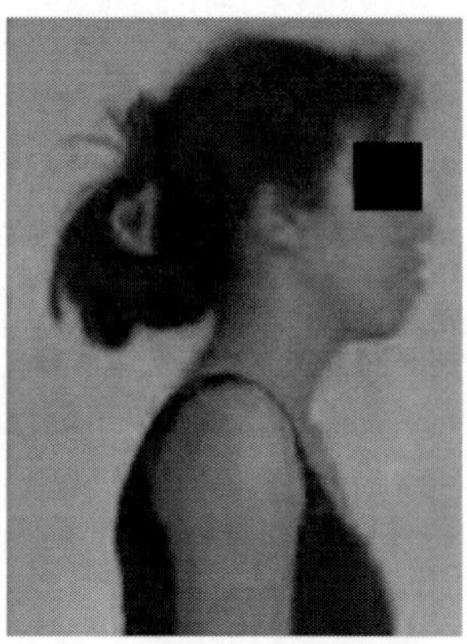

Figure 3. The head forward may be a sign of stress or anxiety.

The effects of emotional tension on posture can be confirmed by the increased electromyographic muscle activity during a condition of mental stress [74] and the increased difficulty of maintaining posture control with the threat caused by increased height. [75] Another study states the there is a relation between physical or psychological stress with neck pain caused by an increased contraction of the upper trapezius, verified by electromyography. [76] It means that stress has an effect on the posture of the shoulder and head. Another author states that anxiety and forward head seems to be related as well [77] (Figure 3).

Experiences with Posture

It is assumed that if the body segments are kept out of alignment for extended periods, the muscles are used in a shortened position and become stronger than their antagonists, which are elongated and weak. [14,78] Strengthening exercises have been prescribed in an attempt to correct postural deviations. In theory, the strengthening of weak and overly stretched muscles results in an adaptive shortening. Together with the stretch of their short antagonists, the body segment would be repositioned. [14,79] However, a number of authors have found no significant association between lumbar lordosis and the strength of the trunk flexors. [80-82] In order to maintain posture for a long time, resistance exercises may be more important than maximal strength exercises. Associations between the resistance of the abdominal muscles and lumbar posture have been sought, but no significant relationship was found. [3,83]

The definition of what classifies as a strong muscle is interesting. A weightlifting champion may have very strong extensors of the trunk with an absence of lower back pain or shortened muscles. It is possible that an investigation into the relationship between flexors and extensors could aid in an improved understanding, although there is difficulty in relating muscle groups with different demands of strength.

Itoi and Sinaki undertook a two-year prospective study to test the hypothesis that increasing the strength of trunk extensors could reduce a hyperkyphosis. [84] The group that performed the trunk extension in prone position ten times, five days per week, increased in strength significantly more than the control group who did not exercise at all. There was no radiographic difference for kyphosis and only a minimal difference for lordosis, which was lower in the group that did exercises. In fact, the trunk extension exercise had a greater effect on the lower back. The eccentric phase of the exercise should have been slower than the positive phase in order to prevent an accident in front of the trunk. There was an emphasis on eccentric muscle contraction, which may have generated a strong stretch of the lower back. No association was found between the horizontal extension range of motion of the shoulder and the abducted scapula. [85] No significant correlation was found between lumbar lordosis and the radiographically measured range of motion in flexion and trunk extension and hip extension in women. [80] There was also no significant association between the length of the back muscles and lumbar lordosis, although there was a significant association between the length of the abdominal muscles and lumbar lordosis. [81]

The effect of the combination of stretching and strengthening on lumbar lordosis has been studied. A small difference in the decrease of lumbar lordosis was found in an experimental group who performed the lumbar spine and stretching exercises for the hip extensor and flexor muscles of the trunk three times a week for a period of four weeks. [87]

The three groups that performed exercises with a vest and electrical stimulation were compared in terms of alterations in idiopathic scoliosis. After three months all groups showed a trend towards a reduction of scoliosis, although there were no significant differences between the groups.

Wang et al. investigated the influence of a program of strengthening and stretching for scapula posture. Exercises were performed three times per week and six weeks after any change in scapular resting posture was noted. [88] In a review of intervention in scoliosis and kyphosis, White and Panjabi questioned the effectiveness of these exercises to correct postural deviations. They suggested that the forces generated by corrective exercises are usually of small amplitude, frequency and duration and thus, are not sufficient to promote permanent changes in muscle length. [89] One possible benefit of an exercise program could be the reeducation of the patient to adopt better posture during daily activities. [90] In a more recent review, Rosário questioned whether resistance training alone actually produces the adaptive shortening of the muscles that causes postural changes. The author concluded that the frequency and duration of exercise programs were ineffective to produce an adaptive shortening of muscles. [91]

It is clear that existing knowledge about posture is insufficient. Moreover, it is not recommended to analyze a single aspect of posture in isolation. For example, it is difficult to assume that lumbar lordosis is only caused by a weakness of the abdominal muscles or short paravertebral muscles.

Other factors may be influencing the adoption of a certain posture. Some of the associated problems include shortening of the diaphragm and other muscles, body image issues and adjustment problems of body segments in relation to the equilibrium point. Techniques involving the *Muscular Chain* theory have been successful in treating postural problems. [91]

The number of studies that have associated posture and emotion is very small. McHugh and Newell [92] reported the possibility of determining the emotions of groups of people, whose faces were hidden, thereby confirming that emotions can only be understood by studying posture and movements. In a similar study, Kret and Gelder [93] instructed the subject to describe the emotion of a single person among a group. Grammer et al. [94] analyzed the movement of a number of subjects and investigated which movements, such as elbow flexion or external rotation for example, were more commonly repeated. The subjects completed a questionnaire in which they reported their emotional state (confidence or irritation) at the time.

Depression, anger, sadness and other emotions have recorded statistical significance when correlated with certain postural deviations. [95] However, until now, there has been no research about the possibility of emotional treatment for postural problems or postural treatment for emotional problems. Exceptions include the empirical studies based on the work of Reich [72] and the contemporary followers of his philosophy. Nevertheless, it seems clear that stretching and strengthening the muscles of a depressed patient, in order to improve their forward and flexed head, will not be enough if this psychological state continues and this same posture is adopted in life.

Conclusion

Posture is a challenging subject to study. It is difficult to assess quantitatively and from a qualitative perspective it can be open to misinterpretations. It is not possible to consider one body segment in isolation, which is one of the most common mistakes in this area. The human body is completely interconnected and relationships in one segment cannot be as reductionist as studying an agonist and its antagonist. The eyes, TMJ and the length of the limbs are, among other factors, all linked to posture. Likewise, it makes no sense to correct a scoliosis without treating the rotations of the lower limb or correcting the difference of the members if any.

Whilst emotions represent further challenges in terms of study, there does appear to be a correlation with postural deviations which should be further investigated. Postural work should aim not only to stretch and strengthen muscles, but also to deal with muscular chains, and to improve associated aspects of the nervous system, as well as factors related to psychological and body image self-perception.

References

[1] Lovejoy OC. The natural history of human gait and posture Part 1 Spine and pelvis. *Gait Posture.* 2005;21:95-112.

[2] Dudeney HE. The Canterbury puzzles and other curious problems. Project: Gutenberg (ebook); 2008.

[3] Rosário JLP, Marques AP, Maluf SA. Aspectos Clínicos do Alongamento: uma revisão de literatura. *Revista Brasileira de Fisioterapia.* 2004;8:83-88.

[4] Liebenson C. Postural correction. *J Bodyw Movement Ther.* 2008; 12:318-9.

[5] Wallden M. The neutral spine principle. *J Bodyw Movement Ther.* 2009; 13:350-61.

[6] Rosário JLP, Diógenes MSB, Mattei R, Leite JR. Can sadness alter the posture? *J Bodyw Movement Ther.* 2013;17:328-31.

[7] James H, Castaneda L, Miller ME, Findley T. Rolfing structural integration treatment of cervical spine dysfunction. *J Bodyw Movement Ther.* 2009;13:229-38.

[8] Darwin CR. The descent of man, and selection in relation to sex. London: John Murray; 1871.

[9] Reich W. The function of the orgasm. New York: Orgone Institute Press; 1942.

[10] Brown R. The authorized summaries of F.M. Alexander's four books. London: STAT Books; 1992.

[11] Bertherat T, Bernstein C. O corpo tem suas razões – Antiginástica e a consciência de si. São Paulo: Martins Fontes; 1987.

[12] Feitis R, Schultz L. Remembering Ida Rolf. North Atlantic Books: Berkeley; 1997.

[13] Myers T. Anatomy trains. 2nd ed. Edinburgh: Elsevier; 2009.

[14] Kendall FP, Mccreary EK, Provance PG, Rodgers M, Romani W. Muscles: testing and function, with posture and pain. Baltimore: Lippincott Williams & Wilkins; 2010.

[15] Feldenkrais M. Consciência pelo movimento. São Paulo: Summus; 1977.

[16] Falconer MA, Mcgeorge M, Begg AC. Obsevations on the course and mechanism os syntom-production in sciatica and low back pain. *J Neurol Neurosurg Psychiatry.* 1948; 2:13.
[17] Jones LH. Spontaneus release by positioning. *Doctor Osteopathy.* 1964;4:109.
[18] Caillet R. Neck and arm pain. Philadelphia: F. A. Davis; 1964.
[19] Caillet R. Soft tissue pain and disability. Philadelphia: F. A. Davis; 1977.
[20] Maitland GD. Peripheral manipulation. London: Butterworth;1970.
[21] Maitland GD. Vertebral manipulation. London: Butterworth;1977.
[22] Magora A. Investigation of the relationship between low back pain and occupation, age, sex, community, education and other factors. *Industr Med Surg.* 1971;39:465-71.
[23] Farfan HF. Mechanical disorders of the low back. Philadelphia: Lea and Febiger; 1973.
[24] Rosário JLP, Sousa A, Cabral CMN, João SMA, Marques AP. Reeducação postural global e alongamento estático segmentar na melhora da flexibilidade, força muscular e amplitude de movimento: um estudo comparativo. *Fisioterapia e Pesquisa.* 2008;15: 12-8.
[25] Sweeting K, Mock M. Gait and posture - assessment in general practice. *Austral Family Physician.* 2007; 36:398-401, 404-5.
[26] Murley GS, Landorf KB, Menz HB, Bird AR. Effect of foot posture, foot orthoses and footwear on lower limb muscle activity during walking and running: a systematic review. *Gait Posture.* 2009;29:172-87.
[27] Bötzel K, Kraft E. Strategies for treatment of gait and posture associated deficits in movement disorders: the impact of deep brain stimulation. *Restorative Neurology Neuroscience.* 2010; 28:115-22.
[28] Grilley P. Working posture and the fascial structures of the lower spine *J Bodyw Movement Ther.* 2008;12:257-66.
[29] Curnow D, Cobbin D, Wyndham J, Choy B. Altered motor control, posture and the Pilates method of exercise prescription. *J Bodyw Movement Ther.* 2009;13:104-11.
[30] Kappler RE. Postural balance and motion patterns. *J Am Osteopathic Assoc.* 1982;81: 69-77.
[31] Lee D. Princípios e práticas da força muscular e das técnicas funcionais. In: Moderna terapia manual da coluna vertebral, Grieve GP. Editorial Médica Panamericana: São Paulo; 1994.
[32] Hoppenfeld S. Propedêutica ortopédica coluna e extremidades. São Paulo: Atheneu. 1980.
[33] Bienfait M. Fáscias e pompagens. São Paulo: Summus Editorial; 1999.
[34] Bienfait M. Fisiologia da terapia manual. São Paulo: Summus Editorial; 1989.
[35] Barral JP. Visceral manipulation II. Seattle: Eastland Press;1988.
[36] Shneerson J. Disorders of ventilation. London: Blackwell Scientific Publications. 1988.
[37] Souchard PE. O stretching global ativo - A reeducação postural global a serviço do esporte. São Paulo: Manole; 1996.
[38] Marques AP. Cadeias musculares – um programa para ensinar avaliação fisioterapêutica global. São Paulo: Manole; 2000.
[39] Gurney B. Leg length discrepancy. *Gait Posture.* 2002;15:195-206.
[40] Souchard PE, Ollier M. As escolioses. São Paulo: É realizações; 2001.
[41] Bricot B. Posturologia, São Paulo: Ícone; 1999.

[42] Huggare JA, Raustia AM. Head posture and cervicovertebral and craniofacial morphology in patients with craniomandibular dysfunction. *J Craniomandib Pract.* 1992;10: 173-9.

[43] Kapandj AI. Fisiología articular. Tronco y raquis. Madrid: Editorial médica panamericana; 1999.

[44] Makofsky HW. The effect of head posture on muscle contact position: the sliding cranium theory. *J Craniomandib Pract.* 1989;7:286-92.

[45] Jewell MJ. Visão geral da estrutura e função do sistema nervoso central. In: Umphred DA, Schmitz TJ. Fisioterapia neurológica. São Paulo: Manole;1994.

[46] Knight KL. Crioterapia no tratamento das lesões esportivas. São Paulo: Manole; 2000.

[47] Ge W, Long CR, Pickar GJ. Vertebral position alters paraspinal muscle spindle responsiveness in the feline spine: effect of positioning duration. *J Physiol.* 2005; 569:2 655-65.

[48] Schmitz TJ. Examination of Sensory Function. In: Umphred DA & Schmitz TJ. Physical rehabilitation. Philadelphia: F.A. Davis; 2007. p.121-158.

[49] Berne RM, Levy MN. Fisiologia. Rio de Janeiro: Editora Guanabara Koogan; 1988.

[50] Bicas HEA. Torcicolo. Posição viciosa de cabeça. *Medicina, Ribeirão Preto.* 2000; 33:64-72.

[51] Halperin E, Cohen BS. Perceptual-motor dysfunction. Stumbling block to rehabilitation. *Maryland Med J.* 1971;20:139.

[52] Rumsey N, Clarke A, Musa M. Altered body image: the psychosocial needs of patients. *Br J Community Nurs.* 2002;7:563-6.

[53] Lamm-Warburg C. Estratégias para avaliação e planejamento do tratamento das deficiências da percepção. In: O'Sullivan SB, Schimitz TJ. Fisioterapia avaliação e tratamento. São Paulo: Manole; 1993. p. 93-120.

[54] Matusevich D, Garcia A, Gutt S, de la Parra I, Finkelsztein C. Hospitalization of patients with anorexia nervosa: a therapeutic proposal. *Eat Weight Disord.* 2002;7:196-201.

[55] Farrington A, Waller G, Neiderman M, Sutton V, Chopping J, Lask B. Dissociation in adolescent girls with anorexia: relationship to comorbid psychopathology. *J Nerv Ment Dis.* 2002; 190:746-51.

[56] Tamburrino MB, McGinnis RA. Anorexia nervosa. A review. *Panminerva Med.* 2002;44:301-11.

[57] Sacks OW. O homem que confundiu sua mulher com um chapéu e outras histórias clínicas. São Paulo: Companhia das letras; 1997.

[58] Damásio A. O Erro de descartes: Emoção, razão E cérebro humano. Lisboa: Publicações Europa-América; 1995.

[59] Levenson RW, Ekman P, Friesen WV. Voluntary facial expression generates emotion-specific nervous system activity. *Psychophysiology.* 1990; 27: 363-84.

[60] Izard CE. The psychology of emotions. New York: Plenum Press; 1991.

[61] Aristóteles. Arte Retórica e Arte Poética. São Paulo, Difusão Européia do Livro. 1959.

[62] Izard CE. Emotion-cognition relationship and human development. In CE Izard, J Kagan, RB Zajonc (Eds.), Emotions, cognition, and behavior. New York: Cambridge University Press; 1984. p.17-37.

[63] Ekman P, Levenson RW, Friesen WV. Autonomic nervous system activity distinguishes between emotions. *Science.* 1983; 221, 1208-10.

[64] Levenson RW, Carstensen LL, Friesen WV, Ekman P. Emotion, physiology, and expression in old age. *Psychol Aging.* 1991; 6, 28-35.
[65] Siegel BS. Amor, Medicina e Milagres - A cura espontânea de doentes graves, Segundo a experiência de um Famoso cirurgião Norte-Americano. São Paulo: Best Seller; 1989.
[66] Frijda NH. Moods, emotion episodes, and emotions. In: Lewis M, Haviland IM (eds). Handbook of emotions. New York: Guilford Press; 1993.p. 381-403.
[67] Darwin CR. A origem das espécies e a seleção natural. Curitiba: Hemus. 1989.
[68] Goldspink G. Sarcomere length during post-natal growth in mammalian muscle fibers. *J Cell Sci.* 1968; 3:539-48.
[69] Williams PE, Goldspink G. Longitudinal growth of striated muscle fibers. *J Cell Sci.* 1971; 9: 751-67.
[70] Williams PE. Use of intermittent stretch in the prevention of serial sarcomere loss in immobilized muscle. *Ann Rheum Dis.* 1990; 49: 316-7.
[71] Darwin CR. The expression of the emotions in man and animals. New York: d. Appleton and company; 1899.
[72] Reich WA função do orgasmo. São Paulo: Brasiliense. 1995.
[73] Keleman S. Embodying experience: Forming a personal life. Berkeley: Center Press. 1987.
[74] Loon E, Van Masters RSW, Ring C, Mcintyre DB. Changes in limb stiffness under conditions of mental stress. *J Mot Behav.* 2001; 33: 153-64.
[75] Adkin AL, Frank JS, Carpenter MG, Peysar GW. Postural control is scaled to level of postural threat. *Gait Posture.* 2000; 12:87-93.
[76] Shahidi B, Haight A, Maluf K. Differential effects of mental concentration and acute psychosocial stress on cervical muscle activity and posture. *J Electromyogr Kinesiol.* 2013;23:1082-9.
[77] Rosario JLP. Avaliação do corpo para tratamento emocional. Ed. Barauna. São Paulo; 2011.
[78] Novak CB, Mackinnon SE. Repetitive use and static postures: a source of nerve compression and pain. *J Hand Ther.* 1997; 10:151-9.
[79] Alter MJ. Ciência da flexibilidade. Artmed. 1996.
[80] Flint MM. Lumbar posture: a study roentogenographic measurement and the influence of flexibility and strenght. *Res Q.* 1962; 34:15-21.
[81] Walker ML, Rothstein SD, Finucane SD, Lamb RL. Relationships between lumbar lordosis, pelvic tilt, and abdominal muscle performance. *Phys Ther.* 1987;67:512-16.
[82] Youdas JW, Garret TR, Harmsen S, Sauman VJ, Carey JR. Lumbar lordosis and pelvic inclination of asymptomatic adults. *Phys Ther.* 1996; 76:1066-81.
[83] Mulhearn S. George K. Abdominal muscle endurance and its association with posture and low back pain. *Physiotherapy.* 1999; 85:210-16.
[84] Itoi E, Sinaki M. Effect of back strengthening exercise on posture in healthy women 49 to 65 years of age. *Mayo Clin Proc.* 1994; 69:1054-9.
[85] Coppock DE. Relationship of tightness of pectoral muscles to round shoulders in college women. *Res Q.* 1958; 29:146-153.
[86] Alizadeh MH. Standring J. The effect of an exercise regime on lumbar spine curve. In: The engineering of sport. Haake S editor. Rotterdam: A.A. Balkemia; 1996.
[87] El-Sayyad M, Conine TA. Effect of exercise, bracing and electrical surface stimulation on idiopathic scoliosis: a preliminary study. *Int J Rehabil Res.* 1994;17:70-4.

[88] Wang CH, Mcclure P, Pratt NE, Nobilini R. Stretching and strengthening exercises: their effect on dimensional scapular kinematics. *Arch Phys Med Rehabil.* 1999;80:923-29.

[89] White AA, Panjabi MM. Practical biomechanics of the spine. 2nd ed. New York: Springer-Verlag; 1990.

[90] Hrysomallis C, Goodman C. A review of resistance exercise and posture realignment. *J Strength Cond Res.* 2001;15:385-90.

[91] Rosário JLP. Efficiency of modified yoga positions to treat postural pathologies associated pain: A literature review. *J Yoga Phys Ther.* 2012;2:128.

[92] McHugh JE, McDonnell R, O'Sullivan C, Newell FN. Perceiving emotion in crowds: the role of dynamic body postures on the perception of emotion in crowded scenes. *Experimental Brain Research.* 2009; 204:361-72.

[93] Kret ME, Gelder B. Social context influences recognition of bodily expressions. *Experimental Brain Research.* 2010; 203:169-180.

[94] Grammer K, Fink B, Oberzaucher E, Atzmüller M, Blantar I, Mitteroecker P. The representation of self reported affect in body posture and body posture simulation. *Collegium Antropologicum.* 2004;28:159-173.

[95] Rosário JLP, Nakashima IY, Rizopoulos K, Kostopoulos D, Marques AP. Improving posture: Comparing segmental stretch and muscular chains therapy. *Clin Chiropractic.* 2012; 15:121:128.

In: Posture: Types, Exercises and Health Effects
Editor: Sarah A. Curran

ISBN: 978-1-63117-252-6

Chapter 5

The Influence of Fear, Happiness and Concern on Posture

José Luís Pimentel do Rosário,[*] *PT, PhD*
Federal University of São Paulo, São Paulo, Brazil

Abstract

Emotion can be described as a complex set of interactions mediated by neurohormonal systems, which can give rise to affective experiences, generate cognitive processes, activate widespread physiological adjustments to elicit conditions, affect behaviour, attention and responses to events. Posture is defined as the way the body segments relate with each other in a determined position, and is a behaviour that has connections with emotions. This chapter presents a study which investigated the existence of a relationship between body posture and subjective fear, concern and happiness in 28 women (range 20 - 39 years) who had a normal body mass index range (or were underweight) and an absence of neurological, psychiatric or musculoskeletal disorders. The degree of each emotion was rated by analogue scales representing *current* and *usual fear*, *concern* and *happiness*. The results indicated that a relationship existed between *current fear* and protrusion of the shoulders (p = 0.05). *Current* and *usual concern* related to angle of tales, both with the same p value (p = 0.02). *Current happiness* is related to inclination of the shoulder (p=0.04 and r=0.387) and *usual happiness* is related to lateral elevation of the shoulder (p=0.05 and r=0.367). However, there was no relationship between *usual fear* and the postural alterations assessed. In conclusion, fear, concern and happiness may lead to postural alterations.

Keywords: Posture, fear, concern, happiness, emotion, postural assessment, analogue scales

[*] Contact email: joerospt@gmail.com

Introduction

Emotions, Fear, Concern and Happiness

Emotion is a complex set of interactions mediated by neurohormonal systems, which can give rise to affective experiences, generate cognitive processes, activate widespread physiological adjustments to elicit conditions and affect behaviour. [1] Emotions can also affect attention and responses to events, as well as activating relevant associative networks in memory. [2] A painful or traumatic experience is encoded in the central nervous system by the amygdala as a non-verbal memory and stored in patterns of muscle contractions and neurovegetative sensations by the autonomic nervous system, which connects emotions to organic functions. [3] It is important to consider that each emotion is the integration of specific neurochemical and motor processes, and it can be considered as the primary motivational system in humans. [2] The emotion of sadness can be seen as a reaction to the failure to achieve or maintain a goal. [4]

Emotion comes from the Latin *emotionem*, which means *in motion*. Thus, when we speak of emotions, as the Greeks and Romans already knew, we talk of what drives us to take an action or not. Fear is an emotion induced by a perceived threat. It is a basic survival mechanism, the ability to recognize danger leading to a fight-or-flight response. [5] All people have an instinctual response to potential danger, which is in fact important to the survival of all species. The reactions elicited from fear are seen through advantages in evolution. [6]

Concern is related to fear and is characterized by a mental attempt to avoid anticipated potential threat, accompanied by a negative emotional charge of worry or anxiety. A moderate amount of worrying may even have positive effects, if it prompts an individual to take precautions or avoid risky behaviours. However, excessive worry is the main component of generalized anxiety disorder.

Happiness is more difficult to define because it has a broad definition and can mean many different things. Many biological, psychological, religious, and philosophical approaches have attempted to define this emotion. From a minimalist point of view, happiness can be described as emotional state of well-being characterized by positive or pleasant emotions ranging from contentment to intense joy. It is an emotion of fundamental importance that acts like a reward for achieving a goal or motivation (pursuit of happiness).

Happiness is not solely derived from external, momentary pleasures. It seems that there are five main sources of happiness for humans: 1. Pleasure (e.g.: tasty food, warm baths.); 2. Engagement (the absorption of an enjoyed yet challenging activity); 3. Relationships (social ties); 4. Meaning (a perceived quest or belonging to something bigger); and 5. Accomplishments (having realized tangible goals). [7]

Posture

Posture and emotions, empirically have a strong connection. However, scientific evidence for a relationship between these two human characteristics has not yet been reported. Emotions affect the muscles in many ways including the following: the effect of anxiety on heart rate and its rhythm generating a tachycardia; the diaphragm and respiratory rate and hyperventilation; and the skeletal muscles in general. Another example is the effect of long breathing to reduce anxiety. [8,9]

The word posture (from the Latin placement) is used to describe how we stand in space, and also emotions: "His sunken posture conveys defeat." Or for a meaning of attitude: "Our posture toward the new government is still the same." It is also important to note that the range of our posture varies with the time of day (we're all shorter in the afternoon, for instance) as well as our inner attitude (fear, eagerness, fatigue). [10]

The study of human posture is relatively new compared to other areas of medical science, and refers to the alignment and maintenance of body segments in certain positions. [11] Some postural deviations may adversely affect muscular efficiency, predispose individuals to pain and pathological musculoskeletal conditions and provoke unaesthetic alterations. [11,12,13,14] The specific focus on posture is very important since it is related to quality of life, a fact that has stimulated interest in different areas.

Wilhelm Reich was a pioneer in correlating body posture and muscle contraction (muscular armor), which in turn is related to emotional aspects. [15] A program could exist in humans, at the level of the nervous system, which establishes a connection between specific emotions and certain muscle movements. This correspondence would be invariable. [16]

Based on a review of the literature, emotional changes can, with proper stimulation, affect any human being and cause muscle changes coherent with the emotional state. In this context, posture can be an important tool in terms of the diagnosis and treatment of emotional problems. However, there are still very few studies in the literature that have investigated this correlation. This chapter presents a study which set out to investigate the existence of a relationship between the posture and the emotions fear, concern and happiness.

Methods

Twenty-eight women, aged between 20 and 39 years, who had a normal body mass indices between 16 to 24.9 kg/m2 were assessed. The women could not be in their menstrual period at the time of the assessment. The exclusion criteria included any psychiatric, neurological or musculoskeletal disorders. The present study received approval from the Human Research Ethics Committee of the Federal University of São Paulo (UNIFESP) under protocol number 1391/05 and the participants signed a statement of informed consent.

The volunteers were subjected to the same assessment protocol which included demographic data (age, weight in kilograms and height in meters). A digital camera (Canon Power Shot A400) was used for the documentation of the subject's standing right lateral view. The image was transferred to an Intel Core 2 Duo computer and the angle of protrusion of the shoulder was examined by Corel Draw, (Figure 1) as described by Munhoz et al. [17] The first step was to draw a parallel line to the ground. Then, another line was drawn perpendicular to the first line, which had the same function as a plumb line. This line was positioned in the photo, at the very back of the heel of the subject. Another line was drawn from this heel point to the most anterior part of the shoulder. The angle between these two lines, with the heel as the fulcrum, revealed the shoulder's protrusion angle. The temperature in the assessment room was kept at a constant 25°C to avoid possible changes in posture.

The following angles were determined from the lateral view (Figure 1):

- **Protrusion of the head (A):** Another line was drawn from the heel point to the intertragic notch. The angle between this and the plumb line, with the heel as the fulcrum, revealed the protrusion angle of the head.
- **Protrusion of the shoulder (B):** Another line was drawn from the heel point to the most anterior part of the shoulder. The angle between this and the plumb line, with the heel as the fulcrum, revealed the shoulder's protrusion angle.
- **Lateral elevation of the shoulder (C):** an exact copy of the line parallel to the floor (in red) was taken to the suprasternal notch. Another line was drawn from this point towards the upper portion of the shoulder. The supra-sternal notch worked as the fulcrum.
- **Lumbar lordosis (D):** a line was drawn from the most posterior part of the calcaneus (fulcrum) going toward the most anterior point of the lumbar spine, around L2 or L3. The angle was obtained with the line plumb line.
- **Extension of knee (E):** the fulcrum was placed on the popliteal crease. A line was drawn from this point toward the most posterior part of the hamstrings and the other to the most posterior part of the gastrocnemius.

The following angles were involved in the anterior view (Figure 1):

- **Inclination of the head (F):** an exact copy of the line parallel to the floor (in red) crossed the center of the lowest pupil. Another line was drawn leaving this point towards the center of the other pupil, with the center of the lower pupil as the fulcrum.
- **Inclination of the shoulders (G):** an exact copy of the line parallel to the floor (in red) crossed the uppermost part of the acromion. Another line was drawn leaving this point towards the same point on the opposite acromion, with the highest acromion as the fulcrum.
- **Elevation of shoulders (H):** the fulcrum was placed on the suprasternal notch. From this point a line was drawn toward the highest part of the acromion on both sides.
- **Angle of tales (I):** It is used as a marker of scoliosis. It is the angle that arises at the level of the last ribs, because from them up the chest begins to expand and from them down the body expands to reach the hip. When the angles of both sides are equal it indicates that there is not a thoracolumbar scoliosis. The greater the difference between them the greater and more severe the scoliosis is.
- **Valgus knee (J):** the fulcrum was placed on the more medialized point of the knees. A line was drawn from this place towards the medial malleolus on both sides. When the knees were not touching each other to generate the fulcrum point, a copy of both lines was brought towards the center until they touched, creating the angle.

Analogue scales were used to assess the degree of subjective depression, [18] which facilitated an assessment of usual depression (a chronic feeling) and current depression (momentary feeling present at the time of assessment). The scales were characterized by a 10 centimeter line, with the words: *"no depression at all"* to the extreme left and *"Utter depression"* to the extreme right, numbered from 0 to 10 (Figure 2). The volunteers were instructed to make a dash at the place of the line that best described their emotional state. The usual depression scale sheet, designed to help the subjects to understand, said *"you always feel"* right above the numbers. The current depression scale sheet reads *"How you are feeling at this exact moment."*

The linear regression technique and the effect of sample size were calculated using the 'Statistical Package for the Social Sciences' (SPSS) software version 20. The level of significance for the linear regression was set at $p < 0.05$. The power of the sample had a small effect if $r = 0.10$ (explained 1% of the total variance), a medium effect if $r = 0.30$ (explained 9% of the total variance) and a large effect if $r = 0.50$ (explained 25% of the total variance).

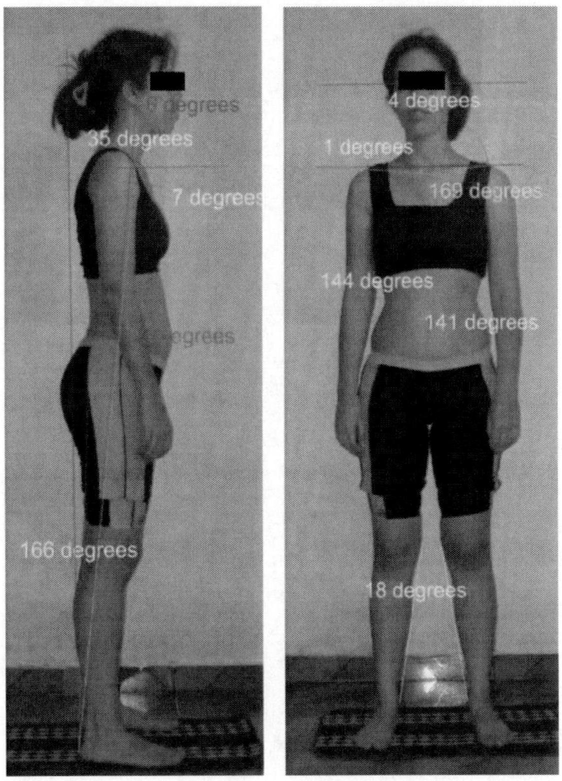

Figure 1. Angle measurements of the lateral and anterior views.

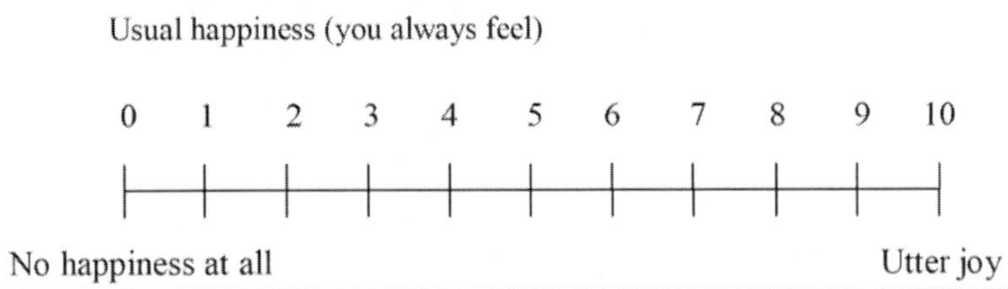

Figure 2. Example of the analogue scales used in the present study. A value of 0 signifies an absence of the emotion studied whereas a value of 10 would be the strongest possible feeling related to that emotion.

Table 1. Correlation index between the degree of self-analyzed depression (analogue scales) both current and usual; and the postural parameters assessed (angle of tales, inclination of the shoulder, protrusion of the shoulder and lateral elevation of the shoulder) with the correlation degree (r) and the statistical significance for each correlation ($p \leq 0.05$)

Postural Variables	Fear				Concern				Happiness			
	Current		Usual		Current		Usual		Current		Usual	
	p	*r*	*p*	*r*	*p*	*r*	*p*	*r*	*p*	*r*	*p*	*r*
Protrusion of the shoulder	0.02	0.431	N.S*		N.S*		N.S*		N.S*		N.S*	
Angle of Tales	N.S*		N.S*		0.02	0.409	0.02	0.430	N.S*		N.S*	
Inclination of the shoulder	N.S*		N.S*		N.S*		N.S*		0.04	0.387	N.S*	
Lateral elevation of the shoulder	N.S*		N.S*		N.S*		N.S*		N.S*		0.05	0.367

*N.S – non-significant values.

Results

Table 1 displays the data obtained from the calculation of the linear regression between the postural and emotional variables. It is possible to note the correlation between *Current fear* and protrusion of the shoulders ($p = 0.05$). The power of the sample value was $r = 0.431$, which indicated a medium sample size effect. *Current* and *usual concern* related to angle of tales, both with the same *p* value ($p = 0.02$) and similar sample size effect (0.409 and 0.430 respectively). *Current happiness* is related to Inclination of the shoulder (p=0.04 and r=0.387) and *usual happiness* is related to lateral elevation of the shoulder (p=0.05 and r=0.367). *Usual fear* did not show a correlation with any of the studied variables. All other postural variables assessed did not show any correlation with the emotions studied.

Discussion

It was possible to confirm a statistically significant association between protrusion of the shoulder and usual subjective sadness. These results are similar to those of other authors who have stated that emotions are related to patterns of contraction of the facial muscles [16,19] and body posture. [20] These results demonstrated that there are postures taken by the body in response to the experience of a specific emotion. There were no statistically significant associations between protrusion of the shoulder and current subjective sadness at the time of testing.

Based on these results it is possible to observe relationships between the variables current happiness and inclination shoulder (p = 0.04). The inclination of the shoulders is a disorder seen in the frontal plane, meaning that one shoulder is higher than the other. It may be related to scoliosis or a tenser trapezius in one side. Since the angle of tales is also an indicator of scoliosis and is not associated with happiness, it makes sense to imagine that this emotion has a connection with a trapezius muscle tension. A relation was found among *usual happiness* the shoulder elevation in lateral view (p = 0.05). As the elevation of the shoulders in the lateral view may remain unilateral, it is possible to be the same alteration found in *current happiness*, shoulder inclination. Still, the possibility of a bilateral elevation of the shoulder cannot be ruled out. However, there is a significant result of this emotion lifting the shoulders, which would indicate the potential for bilateral elevation of the shoulder girdle.

The relationships found for *current concern* had significant results with variable angle of tales (p = 0.02) indicating a scoliosis. The same alteration happened for *usual concern* (p = 0.03) and suggested that worries have a tendency of twisting the body.

Current fear had relations with protrusion of the shoulders (p = 0.02), which shows a pattern of closing like a shell in *self-protection*. The shoulders forward are a component of the fetal position. There were no relations of *usual fear* with any of the postural variables studied. The most likely reason is that *usual fear* is related to a postural trait not studied in this research. Also, the absence of associations between *usual fear* and a postural alteration may have occurred because of the environment where the assessments took place. In a neutral situation, an event that causes fear to the subject may not have as strong an influence at the time of assessment. If this fear was not experienced with the relevant intensity when the photo was taken, this could explain why the posture does not alter significantly in response.

In a previous study by Rosário et al., the protrusion of the shoulders was related to sadness. [21] It is interesting to note how the shoulders are communicative. It would appear that if you are sad, you may have the tendency of protruding the shoulders in an introspective posture. The fear have the same result, however, it would be a protective movement. Even though, both have the same meaning: *"leave me alone!"* Totton and Edmondson [22] stated that a caved chest with the shoulders slumped down and forward is an image of defeat and that people who are stuck in this posture, protecting the heart, have generally given up, probably due to constant frustration.

The same authors suggest some exercises which free the shoulders and can lead to an emotional improvement. In contrast, happiness has the tendency of raising the shoulder in an almost opposite movement. Chinese Medicine states that every emotion is good and bad at the same time and that they are all necessary. However, too much or too few of a specific emotion may cause harm. [23]That is why even happiness, the most desired emotion, in excess can cause postural alterations.

Some postures had no significant relations with any emotion. Although, it does not mean that a relationship does not exist just because it was not found. Lumbar lordosis changes are very difficult to analyze through photos: the volunteers frequently cover part of the lumbar spine with elbows during the picture. This may be the reason why it had no significant results. Other changes involving rotations are very difficult to measure. Rotations can make big differences with small angles, which facilitate mistakes finding the landmarks.

The aim of the present study was not to find an emotional treatment for posture problems or a postural treatment based on cultivating positive emotions. However, since a correlation was detected between emotions and postural deviations, further studies of this treatment are suggested.

Conclusion

A significant association was found in the present study between protraction of the shoulder and current fear. *Current* and *usual concern* related to angle of tales. *Current happiness* is related to inclination of the shoulder while *usual happiness* is related to lateral elevation of the shoulder. No associations were found with *usual fear*.

The findings presented may be useful for clinical practice by helping to identify a patient's emotional condition even without verbalization. They also possibly suggest a new method for treating a sadness component of depression through postural correction, or vice versa, by helping to correct posture in the case of an emotional improvement. Further studies on this topic should be conducted in order to map and trace postural changes related to emotional states and mood disorders, with proper application in clinical practice.

References

[1] Kleinginna PR, Kleinginna AM. A categorized list of emotion definitions with suggestions for a consensual definition. *Motivation Emotion*. 1981;5:345-79.
[2] Izard CE. The psychology of emotions. New York: Plenum Press; 1991.
[3] Schulz ML. Despertando a intuição - usando a sintonia entre mente e corpo para o entendimento e a cura. Objetiva, Rio de Janeiro; 1998.
[4] Stein NL, Trabasso T. The organization of emotional experience: creating links among emotion, thinking and intentional action. *Cognition Emotion*. 1992;6:225-44.
[5] Ohman A. Fear and anxiety: Evolutionary, cognitive, and clinical perspectives. In: Lewis M, Haviland-Jones JM (eds.). Handbook of emotions. New York: The Guilford Press; 2000. p. 573-593.
[6] Olsson A. Phelps EA. Social learning of fear. *Nature Neuroscience*. 2007; 10:1095-102.
[7] Seligman MEP. Can happiness be taught? *Daedalus J.* Spring; 2004.
[8] Darwin CR. The expression of the emotions in man and animals. New York: Appleton and company; 1899.
[9] Garbossa A, Maldaner E, Mortari DM, Biasi J, Leguisamo CP. Effects of physiotherapeutic instructions on anxiety of CABG patients. *Revista Brasileira de Cirurgia Cardiovasc*. 2009;24:359-66.
[10] Myers T. Acture! Posture in action. *Massage & Bodywork magazine*; 2006.
[11] Rosário JLP, Marques AP, Maluf SA. Aspectos clínicos do alongamento: uma revisão de literatura. *Revista Brasileira de Fisioterapia*. 2004;8:83-8.
[12] Liebenson C. Postural correction. *J Bodyw Move Ther.*2008; 12:318-19.
[13] James H, Castaneda L, Miller ME, Findley T. Rolfing structural integration treatment of cervical spine dysfunction. *J Bodyw Move Ther.*2009;13:229-38.
[14] Wallden M. The neutral spine principle. *J Bodyw Move Ther.* 2009;13:350-61.
[15] Reich W. The function of the orgasm. New York: Orgone Institute Press; 1942.
[16] Ekman P, Levenson RWE, Friesen WV. Autonomic nervous system activity distinguishes between emotions. *Science*. 1983;221:1208-10.

[17] Munhoz WC, Marques AP, Siqueira JTT. Evaluation of global body posture in individuals with internal temporomandibular joint derangement. *J Craniomandibular Practice.* 2005;23:1-9.

[18] Williams VSL, Morlock RJ, Feltner D. Psychometric evaluation of a visual analog scale for the assessment of anxiety. *Health Quality Life Outcomes.* 2010; 8:57-63.

[19] Ekman P, Sorenson ER, Friesen WV. Pan-cultural elements in facial displays of emotion. *Science.* 1969;164: 86-8.

[20] Grammer K, Fink B, Oberzaucher E, Atzmüller M, Blantar I, Mitteroecker P. The representation of self reported affect in body posture and body posture simulation. *Collegium Antropologicum.* 2004;28:159-73.

[21] Rosário JLP, Diógenes MSB, Mattei R, Leite JR. Can sadness alter posture? *J Bodyw Move Ther.* 2013;17:328-31.

[22] Totton N, Edmondson E. Reichian growth work – melting the blocks to life and love. Ross-on-Wye: PCCS Books; 2009.

[23] Veith I (translated). The yellow emperor's classic of internal medicine translated by Ilza Veith. Berkley and Los Angeles: University of California Press; 2002.

In: Posture: Types, Exercises and Health Effects
Editor: Sarah A. Curran
ISBN: 978-1-63117-252-6

Chapter 6

Influence of Hamstring Extensibility on Spinal and Pelvic Postures in Highly Trained Athletes

Pedro A. López-Miñarro,* PhD, José M. Muyor, PhD, MSc[2], Fernando Alacid, PhD, MSc[3] and Raquel Vaquero, MSc[4]

[1]Department of Physical Education. University of Murcia
Main Researcher of Physical Exercise and Health Group
[2]Faculty of Education Sciences, Nursing and Physiotherapy,
Laboratory of Kinesiology, Biomechanics and Ergonomics (KIBIOMER),
University of Almería, Almería, Spain
[3]Department of Sport Science at the Catholic University of San Antonio of Murcia,
Murcia, Spain
[4]Sport Traumatology at the Catholic University of San Antonio of Murcia,
Murcia, Spain

Abstract

Hamstring muscle extensibility is an important component of health-related fitness. Because the hamstring muscles originate in the ischial tuberosity of the pelvis, it is logical that tension in the hamstring muscles will have some effect on movement and posture of the pelvis. Some studies have compared the spinal and pelvic postures between subjects in relation to their hamstring extensibility. Reduced hamstring muscle extensibility has been associated with decreased flexion range of motion of the hip and lumbar spine and increased thoracic flexion when trunk flexion movements are performed. However, most of these studies have analyzed non-athlete populations. This chapter presents a study which aimed to compare spinal and pelvic postures in relation to hamstring muscle extensibility in highly-trained athletes. A total of 260 highly trained

* Corresponding author: Email: palopez@um.es

athletes (mean age: 20.56 ± 5.86 years) were recruited. Hamstring muscle extensibility was determined by the passive straight leg raise test (PSLR) in both legs. Thoracic and lumbar curvatures and pelvic position were evaluated with a Spinal Mouse system in standing position and maximal trunk flexion with knees extended (sit-and-reach test) and flexed (Macrae & Wright test position). The sample was divided into three groups according to the straight leg raise angle (Group 1: PSLR < 73°, n=84; Group 2: PSLR ≥ 73° and < 84°, n=90; and Group 3: PSLR ≥ 84°, n=86). The mean values (± SD) of PSLR angle were 63.15 ± 7.01° for Group 1; 77.77 ± 7.41° for Group 2; and 92.06 ± 7.99° for Group 3 (p < 0.001). No significant differences were detected between groups when standing. In the sit-and-reach test, the athletes with lower hamstring extensibility presented significantly higher thoracic angles (Group 1: 65.75 ± 10.83°; Group 2: 61.99 ± 9.57°; Group 3: 57.51 ± 10.81°, p < 0.001) and a more posterior pelvic tilt (Group 1: -16.81 ± 9.88°; Group 2: -12.80 ± 8.62°; Group 3: -6.78 ± 8.35°, p < 0.001). For the Macrae & Wright test, the group with more reduced hamstring extensibility showed higher thoracic angles (Group 1: 59.04 ± 19.54°; Group 2: 44.83 ± 18.29°; Group 3: 42.26 ± 17.19°), lower lumbar flexion (Group 1: 48.72 ± 25.75°; Group 2: 66.02 ± 27.48°; Group 3: 71.62 ± 26.42°) and lower anterior pelvic tilt (Group 1: 52.92 ± 12.42°; Group 2: 58.42 ± 13.81°; Group 3: 59.64 ± 13.03°). Hamstring extensibility has a direct influence on pelvic tilt during flexion movements, especially when knees are extended. The hamstring muscles may influence the spinal and pelvic postures only when the trunk is moderately or maximally flexed and the hamstrings are under tension. Lower hamstring extensibility is related to increased thoracic curve and more posterior pelvic tilt when maximal trunk flexion is performed, which can overload the spine. Since it appears that lower hamstring extensibility is related with poor thoracic and pelvic postures, a systematic stretching programme to improve hamstring extensibility is recommended.

Keywords: thoracic, lumbar, pelvic, spine, straight leg raise, athletes, hamstring, extensibility

Introduction

Sagittal spinal curvatures are geometric parameters which influence mechanical properties. [1, 2] Sagittal alignment influences postural loading and load balance of the intervertebral disc in healthy male and female subjects. [1, 3, 4] Abnormal spinal curvatures cause greater forces to act on the intervertebral discs. Alterations in spinal angles may influence the development of lower back pain. [5, 6]

The sagittal spinal curvatures are considered to influenced by several factors. [7] Hamstring muscle extensibility has been associated with changes in lumbopelvic rhythm [8] and spinal posture. [9-14] Since the hamstring muscles originate at the ischial tuberosity of the pelvis (except for the short head of biceps femoris), the tension in the hamstring muscles influences pelvic posture. [8, 15] The pelvis is considered to be the base for the spine, and its anteroposterior orientation affects the sagittal curves of the spine. For this reason, a change in hamstring extensibility should have some influence in pelvic and spinal posture when the hamstring muscles are subjected to moderate or high tension.

Some studies have suggested that hamstring muscle extensibility influences pelvic and spinal postures in maximal trunk flexion with knees extended. Lower hamstring muscle extensibility has been associated with decreased flexion range of motion of the hip and lumbar spine and increased thoracic flexion. [15-17] Other researchers [18, 19] report an association between greater lumbar spinal flexion and lower hamstring extensibility when the sit-and-reach test is performed.

Some studies have compared the spinal and pelvic postures between subjects in relation to hamstring extensibility. Tully and Stillman [20] observed differences in spinal posture between successful and unsuccessful toe-touchers. Differences in hip (pelvis) flexion between males with or without reduced hamstring extensibility were reported by Gajdosik et al., [16] whilst Carregaro and Coury [21] stated that individuals with decreased hamstring extensibility adopted higher spinal angles and a limitation on pelvic movements during handling tasks. Other studies [22, 23] have suggested differences in pelvic movement in individuals with short and long hamstring during an active straight leg raise. López-Miñarro et al. [24] found that static stretching of the hamstring muscles was associated to immediate changes in the sagittal spinal curvatures and pelvic position. This in turn facilitated greater lumbar flexion and anterior pelvic tilt, as well as lower thoracic kyphosis when performing maximum trunk flexion with the knees extended. Whilst these studies show clear evidence of a relationship between altered spinal and pelvic postures and hamstring extensibility, the data was based on a non-athletic population.

In recent years, several studies have focused on the relationship between hamstring extensibility and spinal posture in athletes. In paddlers, López-Miñarro and Alacid [9] and Muyor et al. [17] found that hamstring muscle extensibility influenced the thoracic and pelvic postures when the sit-and-reach test was performed. In another study, López-Miñarro et al. [25] reported that paddlers with a lower straight leg raise angle adopted a more flexed posture and greater posterior pelvic tilt when sitting in their boat. In cyclists, Muyor et al. [26] observed that hamstring muscle extensibility had a significant relationship in maximal trunk flexion postures with knees flexed and extended. No relationship was noted whilst standing or in the bicycle postures with different handlebars.

Whilst it is evident that a relationship exists between hamstring muscle extensibility, spinal and pelvic posture, there is a need to explore this link in various dynamic positions in an athletic population. Therefore, the aim of this chapter was to present a study that compared spinal and pelvic postures between athletes in relation to their hamstring muscle extensibility.

Material and Methods

A total of 260 highly trained paddlers (mean age: 20.56 ± 5.86 years; height: 172.16 ± 8.70 cm; body mass: 64.01 ± 9.21 kg) were recruited for this study. The inclusion criteria were more than 4 years' paddling experience and training at least six times per week. Paddlers were excluded if they presented with pain induced or exacerbated by the test procedures, injury preventing participation in paddling training before testing, or known structural spinal pathology.

Procedures

An Institutional Ethical Committee approved the study and all subjects or parents signed an informed consent form before participation. The Spinal Mouse system, a hand-held, computer-assisted electromechanical-based device was used to measure sagittal spinal curvatures and pelvic inclination in relaxed standing, maximal trunk flexion with knees extended (sit-and-reach test) and with knees flexed (Macrae & Wright test position). The Spinal Mouse is an electronic computer-aided measuring device, which measures sagittal spinal range of motion and intersegmental angles in a non-invasive way, a so-called surface-based technique. The device is connected radiographically via an analog-digital converter to a standard PC. The Spinal Mouse is a valid and reliable device for global spinal angles. [27-29]

Hamstring muscle extensibility was determined in both legs by the passive straight leg raise test. The measurements were made in a randomized order. No warm-up or stretching exercises were performed by the subjects prior to the test measurements. The subjects were allowed to rest briefly, standing for 5 minutes between measures. All measurements were made during the same testing session and were administered under the same environmental conditions. The measurements lasted a total of 40 minutes. Participants were instructed not to undertake a weight-training session or strenuous exercise the day before testing to ensure consistent test conditions.

Prior to measurements, the principal researcher determined by palpation and marked on the skin surface with a pencil the spinous process of C7 (starting point) and the top of the anal crease (end point). The Spinal Mouse was guided along the midline of the spine (or slightly paravertebrally in particularly thin individuals with prominent processus spinous) starting at the processus spinous of C7 and finishing at the top of the anal crease (approximately S3). For each testing position, the thoracic (T1-2 to T11-12) and lumbar (T12-L1 to the sacrum) spine and the position of the sacrum and the hips (difference between the sacral angle and the vertical) were recorded (Figure 1). In the lumbar curve, negative values corresponded to lumbar lordosis (posterior concavity). With respect to the pelvic position, a value of 0° represented the vertical position. Thus, a greater angle reflected an anterior pelvic tilt while a lower angle (negative values) reflected a posterior pelvic tilt.

Table of measurement

Segment	Upr	Flex	Ext	U-F	U-E	F-E
Th1/2	4	9	9	5	6	0
Th2/3	2	0	2	-2	0	-2
Th3/4	5	2	6	-3	2	-5
Th4/5	7	7	12	0	5	-5
Th5/6	4	9	6	5	1	3
Th6/7	7	12	7	5	-1	5
Th7/8	4	8	12	4	9	-4
Th8/9	3	9	2	6	0	7
Th9/10	10	22	19	13	10	3
Th10/11	-5	-3	-4	2	1	1
Th11/12	-2	8	7	10	8	1
Th12/L1	-8	10	11	17	18	-1
L1/2	-7	4	8	11	15	-4
L2/3	-11	4	3	14	14	0
L3/4	-7	2	2	9	9	0
L4/5	-3	2	0	5	3	2
L5/S1	-6	3	5	9	11	-2
Sac/Hip J.	25	-3	50	-29	24	-53
Thoracic spine	38	82	78	44	40	4
Lumbar spine	-42	24	28	66	70	-4
Incl.	0	38	90	38	90	-52
Length	426	596	599	170	173	-4

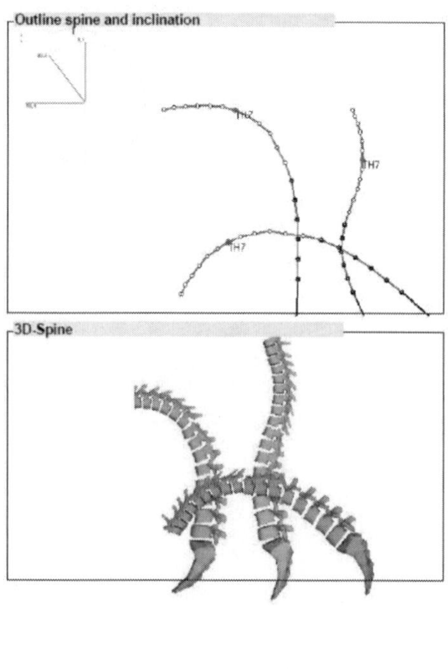

Figure 1. Data obtained from the Spinal Mouse.

Standing

The subject assumed a relaxed position, with the head looking forward, the arms hanging by the side, the knees normally extended, and the feet shoulder-width apart (Figure 2).

Maximal Trunk Flexion with Knees Extended

Spinal angles and pelvic tilt were measured in the sit-and-reach test when the subjects reached the maximal trunk flexion with knees extended. The subjects were required to sit with knees straight and legs together so that the soles of the feet were flat against the end of a constructed box (ACUFLEX I Flexibility tester, height: 32 cm). With palms down, placing one hand on top of the other, the subjects slowly reached forward as far as possible (Figure 3) sliding the hands along the box with the knees as straight as possible and held the position for approximately five seconds while the spinal curvatures and pelvic tilt were measured.

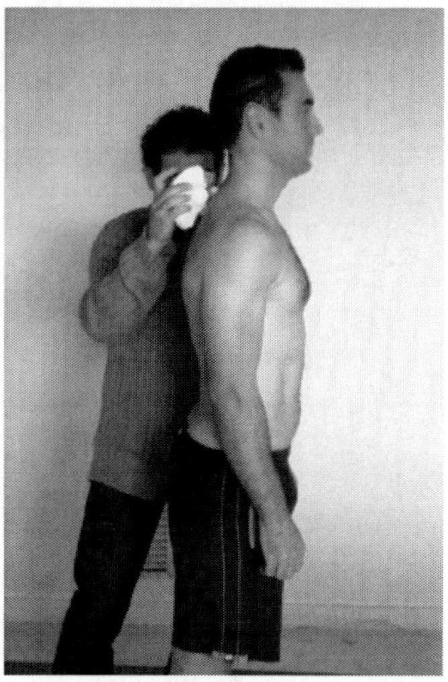

Figure 2. Standing position with the Spinal Mouse applied at the top of the thoracic curve.

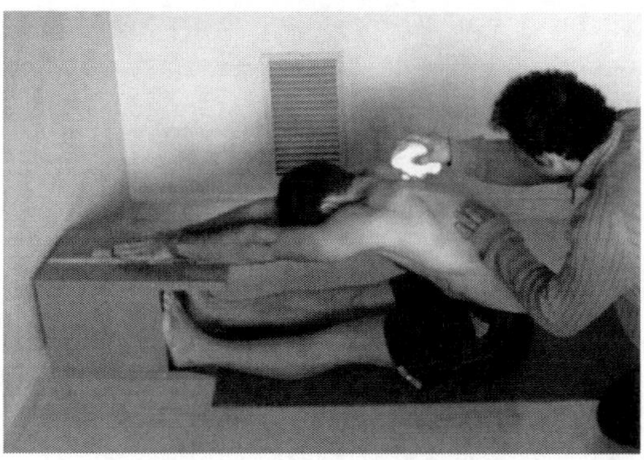

Figure 3. Sit-and-reach test with the Spinal Mouse applied to the thoracic curve.

Maximal Trunk Flexion with Knees Flexed in Sitting

Spinal and pelvic angles were measured when the subjects reached the maximal trunk flexion with knees flexed in sitting position (Figure 4). The subjects sitting with knees flexed (90º) were asked to bend forward maximally. When maximal trunk flexion was achieved the spinal posture was measured.

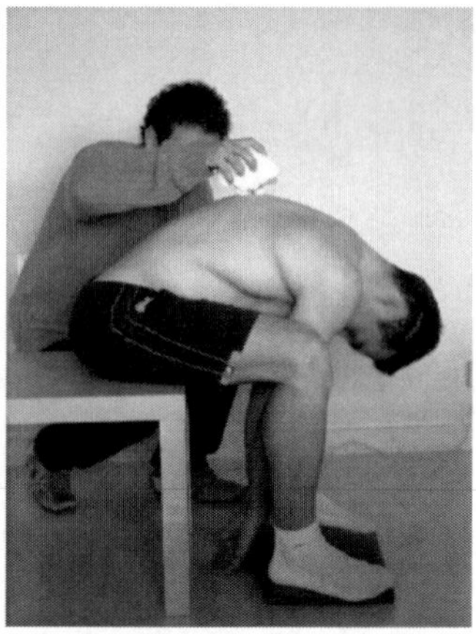

Figure 4. Maximal trunk flexion in sitting with the Spinal Mouse applied in the lower part of thoracic curve.

Hamstring Muscle Extensibility

The criterion measure of hamstring extensibility was determined by performing a passive straight leg raise (PSLR) on each limb in counterbalanced order. While the participant was in the supine position, a Uni-level inclinometer (ISOMED, Inc., Portland, OR) was placed on the distal tibia. The participant's leg was lifted passively by the tester into hip flexion. The knee remained straight during the leg raise. The ankle of the tested leg was restrained in plantar flexion. Moreover, the pelvis was fixed to avoid the posterior pelvic tilt and an auxiliary tester kept the contralateral leg straight to avoid external rotation. [30] The criterion score of hamstring extensibility was the maximum angle (degree) read from the inclinometer at the point of maximum hip flexion. Angles were recorded to the nearest degree for each leg. Two trials were given for each leg and the average of the two trials on each side was used for subsequent analysis.

Statistical Analysis

Only subjects with PSLR difference between right and left sides lower or equal to 5 degrees were included in the analysis. Eighteen subjects were excluded. The left and right PSLR measurements were then averaged. After this, the sample was divided into three groups based on the straight leg raise angle (Group 1: PSLR $< 73°$, n=84; Group 2: PSLR $\geq 73°$ and $<84°$, n=90; and Group 3: PSLR $\geq 84°$, n=86).

The hypotheses of normality and homogeneity of the variance were analyzed via the Kolmogorov-Smirnov and Levene tests, respectively. Descriptive statistics including means and standard deviations were calculated. A one-way ANOVA was used to determine differences among groups for all dependent variables. Significant F-ratios were followed by Tukey´s post hoc analyses to examine pairwise group differences. The data were analyzed using the 'Statistical Package for the Social Sciences' (SPSS) software version 19.0. The level of significance was set at $p \leq 0.05$.

Results

The mean values (± SD) of PSLR angle were 63.15 ± 7.01° for Group 1, 77.77 ± 7.41° for Group 2, and 92.06 ± 7.99° for Group 3 ($p < 0.001$). No significant differences were found between right and leg PSLR angle in any group.

The mean values of thoracic curve, lumbar curve and pelvic tilt for each group are presented in Figures 5, 6 and 7, respectively. No significant differences were detected between groups when standing in any variable. In the sit-and-reach test the athletes with lower hamstring extensibility presented significantly higher thoracic angles (mean ± standard deviation) (Group 1: 65.75 ± 10.83°; Group 2: 61.99 ± 9.57°; Group 3: 57.51 ± 10.81°, $p < 0.001$) and a more posterior pelvic tilt (Group 1: -16.81 ± 9.88°; Group 2: -12.80 ± 8.62°; Group 3: -6.78 ± 8.35°, $p < 0.001$). With respect to the Macrae & Wright test, the group with more reduced hamstring extensibility showed higher thoracic angle (Group 1: 59.04 ± 19.54°; Group 2: 44.83 ± 18.29°; Group 3: 42.26 ± 17.19°), lower lumbar flexion (Group 1: 48.72 ± 25.75°; Group 2: 66.02 ± 27.48°; Group 3: 71.62 ± 26.42°) and a more posterior pelvic tilt (Group 1: 52.92 ± 12.42°; Group 2: 58.42 ± 13.81°; Group 3: 59.64 ± 13.03°).

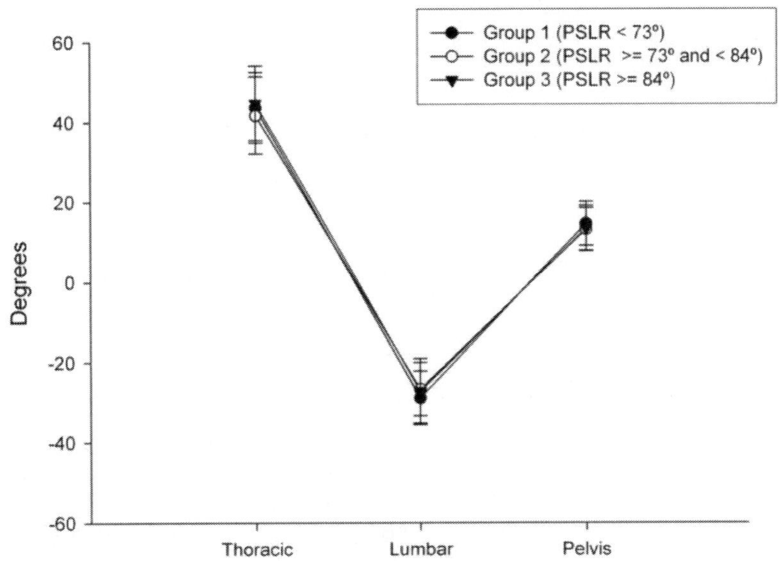

Figure 5. Mean (± standard deviation) of thoracic and lumbar curves and pelvic inclination in standing for the three groups.

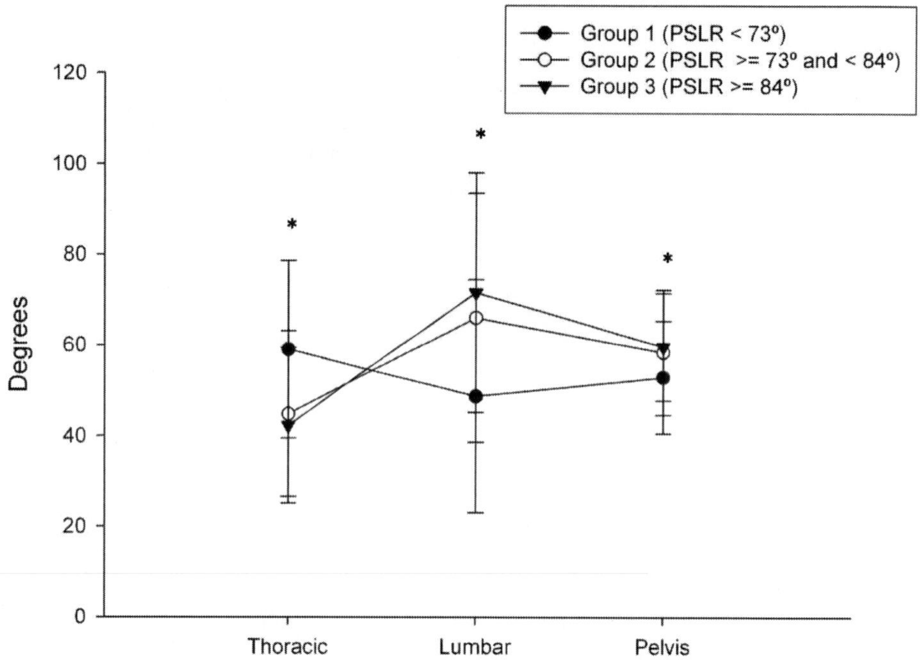

Figure 6. Mean (± standard deviation) of thoracic and lumbar curves and pelvic inclination in maximal trunk flexion with knees flexed for the three groups. * $p < 0.01$ with respect to group 1.

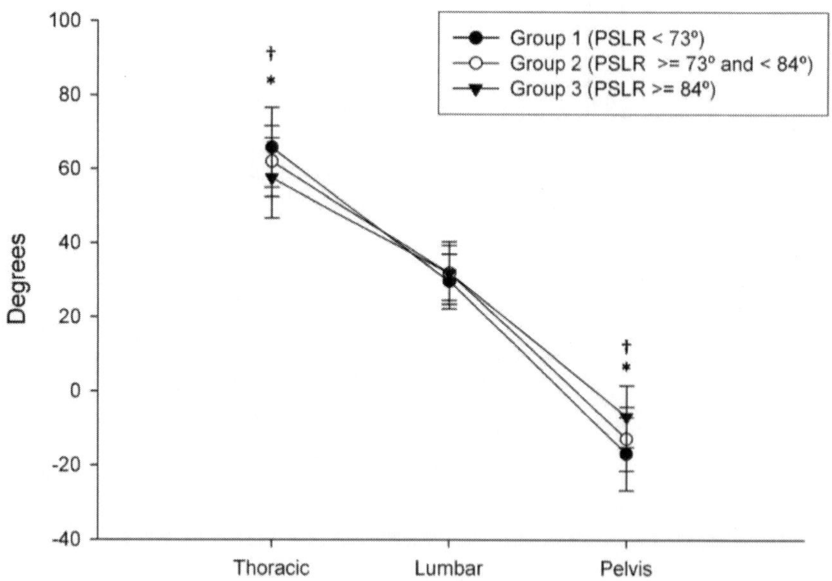

Figure 7. Mean (± standard deviation) of thoracic and lumbar curves and pelvic inclination in maximal trunk flexion with knees extended for the three groups. *$p < 0.01$ between groups 1 and 3; † $p < 0.05$ between groups 2 and 3.

Discussion

The aim of this chapter was to present a study which analyzed the relationship between hamstring muscle extensibility and spinal and pelvic postures in several positions. The main findings showed that lower hamstring extensibility is associated with greater thoracic angles and more posterior pelvic tilt when maximal trunk flexion is performed. However, no differences were found between groups in standing. These results are in agreement with previous studies in younger paddlers, [7, 17] cyclists [26, 31] and young adults. [15] For example, Muyor et al. [26] found a relationship between hamstring extensibility and trunk flexion postures in cyclists. This relationship was stronger when performed a maximal trunk flexion with knees extended. Several studies have found that females reach lower thoracic angles when sit-and-reach tests are performed. [9, 10, 12-14] This posture has been associated to higher hamstring extensibility of females compared to males. [32] However, no association was detected between hamstring extensibility and spinal posture in standing and slumped sitting. In agreement with previous cross-sectional studies [9, 15, 33] the thoracic and lumbar angles and pelvic tilt in standing are not influenced by hamstring extensibility. This may be due to the hamstring muscles are slightly extended, with little passive tension. [7, 34]

Differences between groups were higher in the sit-and-reach test than maximal trunk flexion with knees flexed. The higher thoracic angles during the flexion movements in subjects with lower extensibility could indicate that they were compensating for their pelvic restriction. Tully and Stillman [20] state that subjects with greater capacity to flex the hips with extended knees can reach their toes without needing to use the full available thoracic motion. Decreased hamstring extensibility has been previously associated with poor hip mobility. [35] Hip range of motion influences lumbo-sacral posture due to the muscles' direct attachment to the ischial tuberosities. [15, 21, 22] As the subject bends, the pelvis freely rotates forward until the passive tension in the hamstrings begins to influence pelvic rotation. [34] The hamstring muscles may only influence the spinal and pelvic postures when the trunk is moderately or maximally flexed and the hamstrings are under tension. When the knees are flexed (Macrae & Wright test), the tension in the hamstrings is less and the hamstring extensibility has less influence on pelvic and spinal posture. Indeed, when trunk flexion is performed with knees flexed (90°), the pelvis achieves a larger anterior pelvic tilt.

Hamstring extensibility appears not to affect the lumbar curve in any position. Several studies report a weak and nonsignificant correlation between lumbar curve and hamstring extensibility in bending postures. [12, 36, 37] However, Gajdosik et al. [16] found that lumbar flexion was influenced by hamstring extensibility during maximal trunk flexion, but that it had no effect on standing. In contrast, Kendall et al. [18] and Sahrmann [19] observed an association between excessive lumbar spinal flexion and reduced hamstring extensibility when forward bending or touching the toes.

Previous studies have analyzed the influence of hamstring extensibility in sport specific postures. López-Miñarro et al. [25] evaluated pelvic and spinal postures in kayakers when sitting on their boat. Paddlers with lower extensibility values in the straight leg raise test presented higher thoracic and lumbar flexion and a more posterior pelvic tilt in the kayak in all evaluated positions. However, no significant differences were found between both groups when standing. In a recent study, Kang et al. [38] found a significant increase in hip flexion as well as decreases in both lumbar flexion and the lumbar/hip flexion ratio during the preparation phase of stoop lifting after an acute bout of hamstring-stretching exercises. Hamstring-stretching exercises may be useful for decreasing lumbar flexion and increasing hip flexion in people who perform repetitive stoop lifting in the workplace or during daily activities. [38] López-Miñarro et al. [24] find that a protocol of static stretching of the hamstring muscles was associated to greater lumbar flexion and anterior pelvic tilt as well as lower thoracic kyphosis when performing the sit-and-reach test. However, improved hamstring extensibility had no effect on spinal curvatures and pelvic tilt in either the standing position or maximal trunk flexion in sitting with knees flexed (MacRae & Wright position). Borman et al. [39] found no changes in lumbar flexion range of motion or in lumbar curvature in adults with short hamstring muscles after a 4-week program of hamstring stretching, although hamstring muscle extensibility increased.

The spinal curvatures influence intradiscal pressures, compressive and shear forces in the intervertebral discs. [1-4] Reduced hamstring extensibility is related to increased thoracic angles and posterior pelvic tilt, which can overload the spine during sport and daily activities. The restriction of pelvic movement is considered to be a predisposing factor for low back pain. Thus, if the pelvic tilting is limited, the more lax spinal tissues will be stressed. [40] In fact, it has been recognized that, during trunk flexion, the flexed position of the lumbar spine produces larger shear forces. [41] Greater spinal angles also produce larger contributions from passive components and greater intradiscal pressures on the thoracic and lumbar tissues. This is of special concern in young athletes, because an increased thoracic angle or kyphotic lumbar postures impose great demands on the immature spine, altering the spine's exposure to mechanical loadings during growth.

Stretching can provide a range of health-related motion benefits. Clinicians have generally considered hamstring flexibility training to be an integral component in the prevention and rehabilitation of injuries, as well as a method of improving performance in daily activities and sports. Systematic stretching of the hamstring muscles should be included in the training program of athletes to reduce the thoracic intervertebral flexion and improve anterior pelvic tilt during trunk flexion movements. Li et al. [33] found that hip motion during late and total forward bending was increased after a stretching program. López-Miñarro et al. [24] recommended perform hamstring stretching previously to sport activities involving trunk flexion with the knees straight. Stretching exercises for hamstring muscles should be performed maintaining an appropriate spinal alignment. Some studies have found improvements on hamstring extensibility after a stretching program. [42-44] However, stretching may be beneficial only if the technique employed and the stretch holding times are suitable. Recommendations for duration of hamstrings stretching range from 5 to 60 seconds. Bandy and Irion [45] determined that stretching durations of 30 and 60 seconds resulted in significantly greater range of motion improvements than a 15-second stretch. In the study conducted by Cipriani et al. [46] stretching for 10 seconds 6 times was shown to be equally as effective as stretching 2 times for 30 seconds over 3- and 6-week periods. Roberts and Wilson [47] demonstrated that a 15-second stretch repeated 3 times was more effective than a 5-second stretch repeated 9 times.

Conclusion

The paddlers with lower hamstring extensibility showed a greater thoracic and more posterior pelvic tilt when maximum trunk flexion is performed with knees flexed and extended. Since there appears to be a relationship between lower hamstring extensibility and poor thoracic and pelvic postures, a systematic stretching programme to improved hamstring extensibility is recommended.

References

[1] Keller TS, Colloca CJ, Harrison DE, Harrison DD, Janik TJ. Influence of spine morphology on intervertebral disc loads and stresses in asymptomatic adults: implications for the ideal spine. *Spine J.* 2005;5:297-300.

[2] McGill SM. Low back disorders. Evidence-Based prevention and rehabilitation. Champaign, IL: *Human Kinetics*; 2002.

[3] Polga DJ, Beaubien BP, Kallemeier PM, Schellhas KP, Lee WD, Buttermann GR, et al. Measurement of in vivo intradiscal pressure in healthy thoracic intervertebral discs. *Spine J.* 2004;29:1320-4.

[4] Wilke HJ, Neef P, Hinz B, Seidel H,Claes, LE. Intradiscal pressure together with anthropometric data - a data set for the validation of models. *Clin Biomech.* 2001;1: S111-26.

[5] Harrison DE, Colloca CJ, Harrison DD, Janik TJ, Haas JW, Keller TS. Anterior thoracic posture increases thoracolumbar disc loading. *Eur Spine J.* 2005;14:234-42.

[6] Smith A, O´Sullivan P,Strajer L. Classification of sagittal thoraco-lumbo-pelvic alignment of the adolescence spine in standing and its relationship to low back pain. *Spine J.* 2008;33:2101-7.

[7] López-Miñarro PA, Muyor JM, Alacid F, Rodríguez PL. Influence of sport training on sagittal spinal curvatures. In: Wright AM, Rothenberg SP, editors. *Posture: Types, assessment and control.* New York: Nova Publishers Pub Inc; 2011. P. 63-98.

[8] Esola MA, McClure PW, Fitzgerald GK, Siegler S. Analysis of lumbar spine and hip motion during forward bending in subjects with and without a history of low back pain. *Spine J.* 1996;21:71-8.

[9] López-Miñarro PA, Alacid F. Influence of hamstring muscle extensibility on spinal curvatures in young athletes. *Sci Sports.* 2010;25:188-93.

[10] López-Miñarro PA, Rodríguez PL. Hamstring muscle extensibility influences the criterion-related validity of sit-and-reach and toe-touch tests. *J Strength Cond Res.* 2010;24:1013-8.

[11] López-Miñarro PA, Alacid F, Muyor JM. Comparación del morfotipo raquídeo y extensibilidad isquiosural entre piragüistas y corredores. *Int J Med Sci Phys Act Sport.* 2009;36:379-92.

[12] López-Miñarro PA, Sáinz de Baranda P, Rodríguez-García PL. A comparison of the sit-and-reach test and the back-saver sit-and-reach test in university students. *J Sports Sci Med.* 2009;8:116-22.

[13] López-Miñarro PA, Sáinz de Baranda P, Rodríguez-García PL, Yuste JL. Comparison between sit-and-reach test and V sit-and-reach test in young adults. *Gaz Med Ital.* 2008;167:135-42.

[14] Rodríguez-García PL, López-Miñarro PA, Yuste JL, Sáinz de Baranda P. Comparison of hamstring criterion-related validity, sagittal spinal curvatures, pelvic tilt and score between sit-and-reach and toe-touch tests in athletes. *Med Sport.* 2008;61:11-20.

[15] Gajdosik RL, Albert CR, Mitman JJ. Influence of hamstring length on the standing position and flexion range of motion of the pelvic angle, lumbar angle, and thoracic angle. *JOSPT.* 1994;20:213-9.

[16] Gajdosik RL, Hatcher CK, Whitsell S. Influence of short hamstring muscles on the pelvis and lumbar spine in standing and during the toe-touch test. *Clin Biomech.* 1992;1:38-42.

[17] Muyor JM, Alacid F, Rodríguez-García PL, López-Miñarro PA. Influencia de la extensibilidad isquiosural en la morfología sagital del raquis e inclinación pélvica en deportistas. *Int J Morphol.* 2012;30:176-81.

[18] Kendall FP, McCreary EK, Provance PG, Rodgers MM, Romani WA. Muscles: testing and function with posture and pain (5th ed.). Baltimore: Lippincott Williams & Wilkins; 2005.

[19] Sahrmann SA. Diagnosis and treatment of movement impairment syndromes. St. Louis: Mosby; 2002.

[20] Tully EA, Stillman BC. Computer-aided video analysis of vertebrofemoral motion during toe touching in healthy subjects. *Arch Phys Med Rehabil.* 1997;78:59-66.

[21] Carregaro RL, Coury HJC. Does reduced hamstring flexibility affect trunk and pelvic movement strategies during manual handling? *Int J Ind Ergon.* 2009;39:115-20.

[22] Congdon R, Bohannon R, Tiberio D. Intrinsic and imposed hamstring length influence posterior pelvic rotation during hip flexion. *Clin Biomech.* 2005;20:947-51.

[23] Dewberry MJ, Bohannon RW, Tiberio D, Murray R, Zannotti CM. Pelvic and femoral contributions to bilateral hip flexion by subjects suspended from a bar. *Clin Biomech.* 2003;18:1067-76.

[24] López-Miñarro PA, Muyor JM, Belmonte F, Alacid F. Acute effects of hamstring stretching on sagittal spinal curvatures and pelvic tilt. *J Human Kinet.* 2012;31:69-78.

[25] López-Miñarro PA, Muyor JM, Alacid F. Influence of hamstring extensibility on sagittal spinal curvatures and pelvic tilt in high-trained young kayakers. *Eur J Sports Sci.* 2012;12:469-74.

[26] Muyor JM, López-Miñarro PA, Alacid F. The relationship between hamstring muscle extensibility and spinal postures varies with the degree of knee extension. *J Applied Biomech.* 2013, in press.

[27] Guermazi M, Ghroubi S, Kassis M,Jaziri O, Keskes H, Kessomtini W, et al. Validity and reliability of Spinal Mouse® to assess lumbar flexion. *Ann Readapt Med Phys.* 2006;49:172-7.

[28] Mannion AF, Knecht K, Balaban G, Dvorak J, Grob D. A new skin-surface device for measuring the curvature and global and segmental ranges of motion of the spine: reliability of measurements and comparison with data reviewed from the literature. *Eur Spine J.* 2004;13:122-36.

[29] Post RB, Leferink VJ. Spinal mobility: sagittal range of motion measured with the SpinalMouse, a new non-invasive device. *Arch Orthop Trauma Surg.* 2004;124:187-92.

[30] Santonja Medina FM, Sainz de Baranda Andújar P, Rodríguez García PL, López Miñarro PA, Canteras Jornada M. Effects of frequency of static stretching on straight-leg raise in elementary school children. *J Sports Med Phys Fitness.* 2007;47:304-8.

[31] Muyor JM, López-Miñarro PA, Alacid F. Influence of hamstring muscles extensibility on spinal curvatures and pelvic tilt in highly trained cyclists. *J Hum Kinet.* 2011;29:15-23.

[32] Miñarro PA, Andújar PS, García PL, Toro EO. A comparison of the spine posture among several sit-and-reach test protocols. *J Sci Med Sport.* 2007;10:456-62.

[33] Li Y, McClure PW,Pratt N. The effect of hamstring muscle stretching on standing posture and on lumbar and hip motions during forward bending. *Phys Ther.* 1996;76:836-49.

[34] Shin G, Shu Y, Li Z, Jiang Z, Mirka G. Influence of knee angle and individual flexibility on the flexion-relaxation response of the low back musculature. *J Electromyogr Kinesiol.* 2004;14:485-94.

[35] Sjolie AN. Associations between activities and low back pain in adolescent. *Scand J Med Sci Sports.* 2004;14:352-9.

[36] Hui SS, Yuen PY. Validity of the modified back-saver sit-and-reach test: a comparison with other protocols. *Med Sci Sports Exerc.* 2000;32:1655-9.

[37] Liemohn W, Sharpe GL, Wasserman JF. Criterion related validity of the sit-and-reach test. *J Strength Cond Res.* 1994;8:91-4.

[38] Kang MH, Jung DH, AnDH, Yoo WG, Oh JS. Acute effects of hamstring-stretching exercises on the kinematics of the lumbar spine and hip during stoop lifting. *J Back Musculoskelet Rehabil.* 2013;26:329-36.

[39] Borman NP, Trudelle-Jackson E, Smith SS. Effect of stretch positions on hamstring muscle length, lumbar flexion range of motion, and lumbar curvature in healthy adults. *Physio Theory Pract.* 2011;27:146-54.

[40] Norris CM. Back stability. Champaign, IL: Human Kinetics; 2000.

[41] Potvin JR, McGill SM, Norman RW. Trunk muscle and lumbar ligament contributions to dynamic lifts with varying degrees of trunk flexion. *Spine J.* 1991;16:1099-107.

[42] Nelson RT, Bandy WD. Eccentric training and static stretching improve hamstring flexibility of high school males. *J Athl Train.* 2004;39:254-8.

[43] Reid DA, McNair PJ. Passive force, angle, and stiffness changes after stretching of hamstring muscles. *Med Sci Sports Exerc.* 2004;36:1944-8.

[44] Rodríguez PL, Santonja F, López-Miñarro PA, Sáinz de Baranda P, Yuste JL. Effect of physical education programme on sit-and-reach score in schoolchildren. *Sci Sports.* 2008;23:170-5.

[45] Bandy WD, Irion JM. The effect of time on static stretch on the flexibility of the hamstring muscles. *Phys Ther.*1994;74:845-50.

[46] Cipriani D, Abel B, Pirrwitz D. A comparison of two stretching protocols on hip range of motion: implications for total daily stretch duration. *J Strength Cond Res.* 2003;17:274-8.

[47] Roberts JM, Wilson K. Effect of stretching duration on active and passive range of motion in the lower extremity. *Br J Sports Med.* 1999;33:259-63.

In: Posture: Types, Exercises and Health Effects
Editor: Sarah A. Curran

ISBN: 978-1-63117-252-6

Chapter 7

Spinal Posture in Cycling

José M. Muyor, [1,*] ***PhD, MSc, Pedro A. López-Miñarro,*** [2] ***PhD, Fernando Alacid,*** [3] ***PhD, MSc and Raquel Vaquero-Cristóbal,*** [4] ***MSc***

[1]Faculty of Education Sciences, Nursing and Physiotherapy,
Laboratory of Kinesiology, Biomechanics and Ergonomics (KIBIOMER Lab)
University of Almería, Almería, Spain

[2]Department of Physical Education, University of Murcia, Murcia, Spain
Main Researcher of Physical Exercise and Health Group

[3]Department of Sport Science at the Catholic University of San Antonio of Murcia,
Murcia, Spain

[4]Sport Traumatology at the Catholic University of San Antonio of Murcia,
Murcia, Spain

Abstract

Currently, cycling is a popular sport, in spite of its demanding physical requirements, the interaction with motor vehicles and the need to adopt an unnatural position on the bicycle for better aerodynamics. The analysis of posture in cyclists on the bicycle and their spinal adaptations in different positions has been studied in the field of sports medicine. This chapter describes the results of major research on the position adopted in cycling, the spinal adaptations generated by the systematic practice of this sport, and the variables that influence the posture of cyclists. The studies reviewed suggest that the morphology of the thoracic curve of the cyclist is characterized by a standing hyperkyphotic posture. This condition, however, would be more associated with a lack of postural awareness than with the position on the bicycle because thoracic kyphosis significantly decreases due to the support of the hands on the handlebars. Instead, the lumbar spine modifies its lordosis in the standing position toward a lumbar kyphosis (inversion) on the bicycle, with greater lumbar flexion, as the grip on the handlebars is lower and farther away relative to the seat. This posture, held on the bicycle, has been associated with the ability of the cyclist to adopt greater lumbar flexion in positions of

[*] Corresponding author: Email: josemuyor@ual.es

maximum trunk flexion with bent or extended knees compared with subjects who do not practice this sport. It has also been observed that the degree of hamstring extensibility does not influence the position of the spine and pelvis on the bicycle even though it does influence the spinal position in movements of maximum trunk flexion with extended knees. In conclusion, cycling posture is characterized by maintaining better alignment of the thoracic spine on the bicycle than in the standing position. In contrast, the lumbar spine maintains an inverted position, which leads to spinal adaptations in positions of maximum trunk flexion and could possibly cause low back pain as a result. Due to the characteristics and specific posture requirements of this sport, core strengthening exercises and improvements in the body schema of cyclists are recommended.

Keywords: Cycling, spine, posture, pelvis, core strengthening, lumbar, kyphosis

Introduction

In the standing position, the spine has different physiological curves in the sagittal plane (cervical lordosis, thoracic kyphosis and lumbar lordosis) that are determined by the morphology of the vertebral bodies, the functionality of the intervertebral discs and ligamentous structures, and the anatomical-physiological muscle integrity. [1, 2] Sports practice involves, in many cases, adopting postures considered to be unnatural to execute a precise technical movement or to maintain a particular body position in pursuit of increased athletic performance. Systematically maintaining these postures for sustained periods has been associated with adaptations of sagittal spine alignment. [3] For example, a greater degree of thoracic kyphosis has been found in athletes who participate in sports in which optimum technique is characterized by the maintenance of trunk flexion postures, including skiing, rowing or wrestling, compared with subjects who do not practice these sports. [4-6] However, positions that hold the spine in flexion have been associated with increased intradiscal pressure in the thoracic [7] and lumbar spine, [8] greater distortion of the spinal tissues [9] and high spinal stress; [10] all of which increase the likelihood of suffering, pain, and/or spinal injury.

Currently, cycling is one of the most popular sports both from a competitive standpoint and for amateurs and hobbyists, even though it is considered as a discipline in which several different factors are involved in performance, including high physical demands, weather conditions and interactions with motor vehicles. [11] The predominant position is the seated position on the bicycle with an anterior inclination of the trunk and lumbar flexion to reach the handlebars with the hands. Because this position is considered unnatural, [12] several studies have analyzed the position adopted by the spine during cycling, how different variables affect the spinal position, and whether cycling modifies spinal curvatures in the sagittal plane.

Cycling and Low Back Pain

Low back pain is one of the most common overuse injuries in cycling. [13] Clarsen et al. [14] interviewed 7 professional teams with a total of 116 cyclists and found that 45% had suffered low back pain in the past 12 months and that 41% had received medical care for the pain. Similar research has shown that bicycle fit, improper equipment, training errors, and individual anatomical factors are important variables affecting the occurrence of low back pain in these athletes. [15]

Some authors have tried to explain how changes in different bicycle components would affect the posture of the cyclist and either cause or decrease low back pain. In this regard, Salai et al. [16] analyzed the effect of changing the seat tilt on the incidence of low back pain in recreational cyclists. The authors found that the incidence and severity of low back pain was significantly reduced with an anterior tilt of the seat between 10°-15°. In comparison, Fanucci et al. [17] reported that the incidence and importance of low back pain in cyclists could be reduced if the bicycle's crankset axis was placed behind the seat axis. This arrangement allowed riders to maintain the lumbar spine in a more physiological position compared with the classical design with the bicycle's crankset axis in front of the seat axis.

Although the design of different prototype bicycles and the analysis of how modification of their components affect low back pain or cyclist's posture may be of scientific interest, the reality is that the Union Cycliste Internationale (UCI) currently sets forth regulations that specify the dimensions and characteristics of bicycles that can be certified and therefore used in official cycling competitions. Thus, several authors have evaluated the spinal morphotypes and the postures adopted by cyclists on bicycles certified for training and/or competition.

Spinal Posture on the Bicycle

Few studies in the scientific literature have evaluated the position of the cervical spine in cyclists. Kolehmainen et al. [1] analyzed the influence of handlebar height on the position of the cervical and thoracic spine using a cycle ergometer in 3 handlebar positions (high, medium and low). The results showed that there were no major changes in thoracic kyphosis and lumbar lordosis in any of the handlebar grips evaluated. However, with a low grip, there was cervical spine extension that tripled load moments compared with the vertically aligned position. These authors recommended the use of a high grip on the handlebar, particularly if the cyclist experiences discomfort in the cervical spine.

Usabiaga et al. [18] conducted a radiological and electromyographic study to evaluate changes in the lumbar spine with different handlebar grips using cyclists' own racing bicycles. The authors found that the lumbar spine changed from lordosis in the standing position to kyphosis in the seated cycling position. Low back muscle contraction was proportional to pedaling intensity and decreased with improved aerodynamic positioning. The tone of the thoracic paravertebral muscles depended on the degree of cervical hyperextension. However, the abdominal muscles remained relaxed for all pedaling positions and intensities. The authors concluded that observed changes in the lumbar spine could modify the normal biomechanics of this spinal region, although mechanical load would be reduced by the weight distribution of the trunk on the handlebar grip. Moreover, the imbalance found between the activity of the flexor and extensor muscles of the trunk could cause low back pain in cyclists without adequate physical preparation.

Burnett et al. [19] evaluated whether there were differences between spine kinematics and muscle activity in cyclists with and without low back pain. The results showed that cyclists with nonspecific low back pain suffered a motor control disorder of the lumbo-pelvic region, with greater flexion and rotation of the lower lumbar spine (L3-S2) and a loss of lumbar muscle contraction (multifidus) compared with the asymptomatic group of cyclists. Recently, Van Hoof et al. [20] found similar results when comparing lumbar spine kinematics between groups of cyclists with and without non-specific low back pain. These authors found that cyclists with low back pain suffered a motor control disorder, resulting in increased lumbar flexion during pedaling that was related to a significant increase in pain in that region. Similarly, Srinivasan and Balasubramanian [21] found that cyclists with low back pain had significantly greater muscle fatigue in the medial trapezius and spinal erector muscles than asymptomatic cyclists. These authors concluded that cycling could aggravate low back pain due to increased fatigue in dorsal muscles.

Other studies have evaluated the posture of healthy cyclists on the bicycle in different sports categories. In this regard, Muyor et al. [22] evaluated the sagittal alignment of the thoracic spine in elite cyclists riding their own bicycles compared with subjects who did not practice any sport (Figure 1). The authors observed less thoracic kyphosis when using a high grip on the handlebars as well as less kyphosis compared with the group of non-athletes. They also found that the thoracic spine on the bicycle had angular values similar to the corrected thoracic position. That is, on the bicycle, the thoracic spine was corrected due to the support of the hands on the handlebars, which caused a scapular retropulsion and straightening of the thoracic curve (Figure 2). In contrast, the group of non-athletes had greater thoracic spinal flexion on the bicycle compared with the self-corrected position. The authors explained this finding as the result of adaptation by the cyclist to an aerodynamic position, causing alignment of the thoracic spine to decrease its front surface area and therefore air friction. On the other hand, elite cyclists were found to have significantly more lumbar flexion on the bicycle compared with non-athletes. Furthermore, there was greater lumbar flexion when the grip on the handlebars was lower and further away relative to the seat. [23]

Figure 1. Cyclist´s posture on the bicycle.

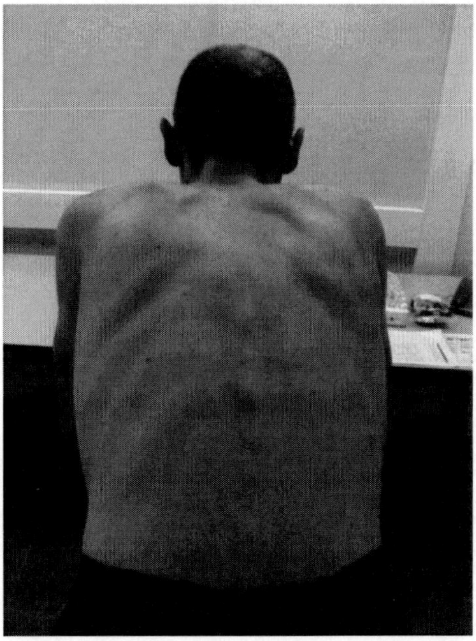

Figure 2. Cyclist´s scapular retropulsion maintained to support of the hands on the handlebars.

Muyor et al. [24] compared spinal postures and pelvic tilts on the bicycle between elite cyclists and Masters 30 category cyclists. The authors found that elite cyclists had lower thoracic flexion and increased lumbar flexion and pelvic tilt compared with Masters 30 cyclists. The differences in the postures adopted on the bicycle were related to age. That is, the younger cyclists (elite) had less thoracic kyphosis in the standing position than the older cyclists (Masters 30). The elite cyclists therefore maintained lower thoracic spine angle values on the bicycle.

Spine Morphology of the Cyclist in the Standing and Trunk Flexion Positions

Several studies have analyzed the spine morphologies of cyclists in different sport categories to determine whether cycling alters the physiological curvature of the spine in the standing position or if there are spinal adaptations in trunk flexion movements. Rajabi et al. [25] compared the thoracic spine angular values in the standing position in a group of cyclists with a minimum cycling training experience of 5 years and a group of non-athletes. The results showed significantly greater thoracic kyphosis in the cyclists compared with the non-athletes. The authors attributed these findings to the specific posture of the cyclist on the bicycle, which significantly increased stress on the spine and thereby increased thoracic kyphosis.

In this regard, Muyor et al. [22] also found significantly higher thoracic angular values in elite cyclists compared with non-athletes in the standing position. However, these authors, unlike Rajabi et al., [25] also evaluated the thoracic spine position on the bicycle and found that both the cyclists and the non-athletes showed a significant decrease in thoracic curvature on the bicycle compared with the standing position. Similar results with respect to decreased thoracic kyphosis on the bicycle have been found in other sport categories in Masters 30 [24] and Masters 40 cyclists. [26]

The categories proposed by Mejia et al. [27] are frequently used to classify thoracic kyphosis: neutral thoracic kyphosis is between 20° and 45°; thoracic hyperkyphosis is > 45°; and thoracic hypokyphosis is < 20°. A high percentage of cyclists have thoracic hyperkyphosis in the relaxed standing position (Figure 3). Specifically, 58.3% of elite cyclists, 53.3% of Masters 30 cyclists [24] and 68% of Masters 40 cyclists. [26] Researchers have suggested that the high percentage of thoracic hyperkyphosis found in cyclists is due to factors such as an imbalance in the body schema or postural attitude rather than the position taken on the bicycle, where the spine is better aligned than in the standing position and therefore is not in a kyphosed position. Later, Muyor et al., [28] after evaluating spinal morphology in the standing position in different age groups of cyclists with the same number of years of cycling practice and training volume experience, found a significant increase in thoracic kyphosis with increasing age in cyclists. These authors concluded that age was a major factor in the angular alignment of the thoracic spine in these athletes.

Similarly, the lumbar spine has also been studied in cyclists. Maintaining an inverted lumbar spine position on the bicycle can cause spinal adaptations in trunk flexion positions. In this sense, Muyor et al. [29] found a significantly higher capacity for lumbar flexion in elite cyclists compared with non-athletes in postures of maximum trunk flexion in seated positions with bent knees and with extended legs. In contrast, no significant differences were found when comparing lumbar flexion in seated positions between elite and Masters 30 cyclists. [30]

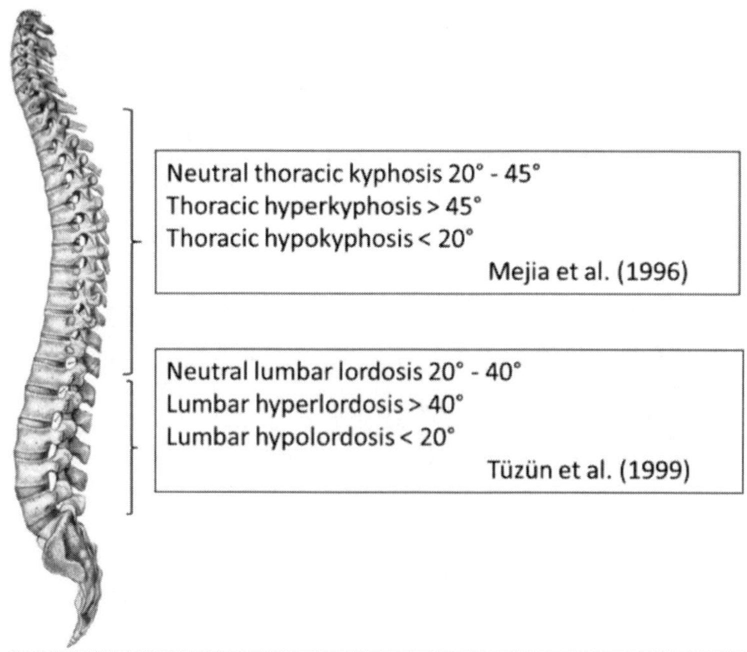

Figure 3. Angle references for spinal curvatures in standing posture.

As discussed earlier in this chapter, a higher frequency of thoracic hyperkyphosis has been found in cyclists in the standing position. The classification proposed by Tüzün et al. [31] is frequently used to classify lumbar spine morphology: neutral lumbar lordosis is between 20° and 40°; lumbar lordosis is > 40°; and lumbar hypolordosis is < 20°. (Figure 3) Most cyclists have a neutral lumbar lordosis in the standing position. This is true for 88.3% of elite cyclists, 76.7% of Masters 30 cyclists, [24] and 68% of Masters 40 cyclists. [26]

Regarding the pelvis, McEvoy et al. [32] compared the degree of the anterior pelvic tilt in the seated position with extended knees (long sitting posture) in a group of elite cyclists with another group of non-athletes. The authors found that cyclists had a significantly greater anterior pelvic tilt than non-athletes. The authors attributed their results to greater adaptation to trunk flexion postures by pelvic tilt in cyclists because that is their normal position on the bicycle. However, these authors did not assess the degree of hamstring extensibility in both groups.

Other studies on athletes of other sports [33] as well as on cyclists [34, 35] have found that the degree of hamstring extensibility influences pelvic movement in positions of maximum trunk flexion with extended knees. Since the hamstring muscles have their anatomical attachment on the ischial tuberosity, when maximal trunk flexion with the knees extended is performed, the pelvis rotates forward until the passive tension in the hamstrings increase significantly. However, in cyclists, the degree of hamstring extensibility does not influence spinal posture [33] or pelvic tilt [36] adopted on the bicycle. This difference is most likely because cyclists' knees never reach their maximum extension on the bicycle during pedaling and therefore the degree of hamstring extensibility would have no effect on the spinal posture adopted on the bicycle.

Core Stability in Cycling

"Core" stability has been defined as the ability to control the position and movement of the trunk on the pelvis, allowing an optimum production, transfer and control of force and motion in the integration of athletic activities. [37] Spine stability refers to the combined and coordinated work of the passive components of the spinal column (bones and ligaments), of the active components (spinal and abdominal muscles), and neuromuscular control (spinal cord and central nervous system).

Asplund and Ross [38] have suggested that core stability is critical for the efficient transfer of power from the lower to the upper extremities and vice versa. Less energy is lost in movement with good core training, which enables the trunk to be more stable, avoiding flexion or rotation movements. Additionally, having a trained core allows cyclists to maintaining their trunk alignment longer, thereby delaying postural fatigue.

In the same vein, core strengthening has also been used as a therapeutic exercise in the treatment of low back pain. [39] In this regard, some studies have reported that the position on the bicycle is related to exercise intensity, ventilatory thresholds, the power capacity, and the performance achieved. [36, 40, 41] Therefore, core training could improve both the posture on the bicycle and prevent the low back pain suffered by these athletes after constantly maintaining the lumbar spine in flexion.

Conclusion

The cyclist's spinal column morphotype in the standing position is characterized by thoracic hyperkyphosis, which is more related to alterations in the body schema than to the position maintained on the bicycle. On the bicycle, the thoracic spine decreases its kyphosis due to the support of the hands on the handlebars and retropulsion of the scapulohumeral girdle. Instead, the lumbar spine changes from lumbar lordosis in the standing position to lumbar kyphosis (inversion) on the bicycle, with greater lumbar flexion when the grip on the handlebar is lower and farther away from the seat.

Whilst maintaining an inverted lumbar spine on the bicycle does not generate alterations in lumbar lordosis in the standing position, it predisposes to greater lumbar flexion in postures of maximum trunk flexion. The degree of hamstring extensibility in cyclists influences the anterior pelvic tilt in positions of maximum trunk flexion with extended knees but has no influence on the spinal position adopted on the bicycle. Due to the characteristics and specific postural requirements of this sport, exercises to strengthen the core and improve the body schema in cyclists are recommended.

References

[1] Kolehmainen I, Harms-Ringdahl K, Lanshammar H. Cervical spine positions and load moments during bicycling with different handlebar position. *Clin Biomech.* 1989;4:105-10.

[2] White AA, Panjabi MM. Clinical biomechanics of the spine. 2 ed. Philadelphia: J. B. Lippincott Company; 1990.

[3] Wojtys EM, Ashton-Miller JA, Huston LJ, Moga PJ. The association between athletic training time and the sagittal curvature of the immature spine. *Am J Sports Medicine.* 2000;28:490-8.

[4] Alricsson M, Werner S. Young elite cross-country skiers and low back pain. A 5-year study. *Phys Ther Sport.* 2006;7:181-4.

[5] *Rajabi R, Doherty P, Goodarzi M, Hemayattalab R.* Comparison of thoracic kyphosis in two groups of elite Greco-Roman and freestyle wrestlers and a group of non-athletic participants. *Brit J Sports Med.* 2008;42:229-32.

[6] Stuchfield BM, Coleman S. The relationships between hamstring flexibility, lumbar flexion, and low back pain in rowers. *Euro J Sports Science.* 2006;6:255-60.

[7] Polga DJ, Beaubien BP, Kallemeier PM, Schelhas KP, Lee WD, Buttermann GR, et al. Measurement of *in vivo* intradiscal pressure in healthy thoracic intervertebral disc. *Spine.* 2004;29:1320-4.

[8] Wilke HJ, Neef P, Caimi M, Hoogland T, Claes LE. New in vivo measurements of pressures in the intervertebral disc in daily life. *Spine.* 1999;24:755-62.

[9] Solomonow M, Zhou BH, Baratta RV, Burger E. Biomechanics and electromyography of a cumulative lumbar disorder: response to static flexion. *Clin Biomech.* 2003;18: 890-8.

[10] Beach TA, Parkinson RJ, Stothart JP, Callaghan JP. Effects of prolonged sitting on the passive flexion stiffness of the in vivo lumbar spine. *Spine J.* 2005;5:145-54.

[11] Winters M, Davidson G, Kao D, Teschke K. Motivators and deterrents of bicycing: comparing influences on decisions to ride. *Transportation.* 2011;38:153-68.

[12] de Vey Mestdagh K. Personal perspective: in search of an optimum cycling posture. *Applied Ergonomics.* 1998;29:325-34.

[13] Marsden M, Schwellnus M. Lower back pain in cyclists: A review of epidemiology, pathomechanics and risk factors. *Int Sport Med J.* 2010;11:216-25.

[14] Clarsen B, Krosshaug T, Bahr R. Overuse injuries in professional road cyclists. *Am J Sports Med.* 2010;38:2494-501.

[15] Asplund C, Webb C, Barkdull T. Neck and back pain in bicycling. *Current Sports Med Reports.* 2005;4:271-4.

[16] Salai M, Brosh T, Blankstein A, Oran A, Chechik A. Effect of changing the saddle angle on the incidence of low back pain in recreational bicyclists. *Brit Sports Med.* 1999;33:398-400.

[17] Fanucci E, Masala S, Fasoli F, Cammarata R, Squillaci E, Simonetti G. Cineradiographic study of spine during cycling: effects of changing the pedal unit position on the dorso-lumbar spine angle. *Radiologia Medica.* 2002;104:472-6.

[18] Usabiaga J, Crespo R, Iza I, Aramendi J, Terrados N, Poza JJ. Adaptation of the lumbar spine to different positions in bicycle racing. *Spine.* 1997;22:1965-9.

[19] Burnett AF, Cornelius MW, Dankaerts W, O´Sullivan PB. Spinal kinematics and trunk muscle activity in cyclists: a comparison between healthy controls and non-specific chronic low back pain subjects -a pilot investigation. *Manual Therapy.* 2004;9:211.

[20] Van Hoof W, Volkaerts K, O´Sullivan K, Verschueren S, Dankaerts W. Comparing lower lumbar kinematics in cyclists with low back pain (flexion pattern) versus asymptomatic controls - field study using a wireless posture monitoring system. *Manual Therapy.* 2012;17:312-7.

[21] Srinivasan J, Balasubramanian V. Low back pain and muscle fatigue due to road cycling - An sEMG study. *J Bodyw Move Ther.* 2007;11:260-6.

[22] Muyor JM, López-Miñarro PA, Alacid F. A comparison of the thoracic spine in the sagittal plane between elite cyclists and non-athlete subjects. *J Back Musculo Rehab.* 2011;24:129-35.

[23] Muyor JM, López-Miñarro PA, Alacid F. Comparison of sagittal lumbar curvature between elite cyclists and non-athletes. *Science & Sport*s, in press.

[24] Muyor JM, López-Miñarro PA, Alacid F. Spinal posture of thoracic and lumbar spine and pelvic tilt in highly trained cyclists. *J Sports Science Med.* 2011;10:355-61.

[25] Rajabi R, Freemont A, Doherty P. The investigation of cycling position on thoracic spine. A novel method of measuring thoracic kyphosis in the standing position. *Arch Physiology Biochemistry.* 2000;1:142.

[26] Muyor JM, Alacid F, López-Miñarro PA. Spinal posture of thoracic and lumbar spine in master 40 cyclists. *Int J Morphology.* 2011;29:727-32.

[27] Mejia EA, Hennrikus WL, Schwend RM, Emans JB. A prospective evaluation of idiopathic left thoracic scoliosis with magnetic resonance imaging. *J Pediatric Orthopaedics.* 1996;16:354-8.

[28] Muyor JM, Alacid F, López-Miñarro PA, Casimiro AJ. Evolution of spinal morphology and pelvic tilt in cyclists of different ages. A cross sectional study. *Int J Morphology.* 2012;30:199-204.

[29] Muyor JM, López-Miñarro PA, Alacid F. Sagittal lumbar curvature in elite cyclists and non-athletic subjects. *Revista Internacional de Medicina y Ciencias de la Actividad Física y del Deporte.* 2012;12:219-31.

[30] Muyor JM, López-Miñarro PA, Alacid F, Casimiro AJ. Sagittal spinal curvatures and pelvic tilt in cyclists: A comparison between two master cyclists categories. *Int SportMed J.* 2012;13:122-32.

[31] Tüzün Ç, Yorulmaz I, Cindaş A, Vatan S. Low back pain and posture. *Clin Rheumatology.* 1999;18:308-12.

[32] McEvoy M, Wilkie K, Williams MT. Anterior pelvic tilt in elite cyclists - A comparative matched pairs study. *Phys Ther Sport.* 2007;8:22-9.

[33] Muyor JM, Alacid F, López-Miñarro PA. Influence of hamstring muscles extensibility on spinal curvatures and pelvic tilt in highly trained cyclists. *J Human Kinetics.* 2011;29:15-23.

[34] Muyor JM, Alacid F, Rodríguez-García PL, López-Miñarro PA. Influence of hamstring extensibility on sagittal spinal curvatures and pelvic inclination in athletes. *Int J Morphology.* 2012;30:176-81.

[35] Muyor JM, López-Miñarro PA, Alacid F. The relationship between hamstring muscle extensibility and spinal postures varies with the degree of knee extension. *J Appl Biomechanics.* 2013, in press.

[36] Sauer JL, Potter JJ, Weisshaar CL, Ploeg HL, Thelen DG. Influence of gender, power, and hand position on pelvic motion during seated cycling. *Med Sci Sports Exer.* 2007;39:2204-11.

[37] Kibler WB, Press J, Sciascia A. The role of core stability in athletic function. *Sports Medicine.* 2006;36:189-98.

[38] Asplund C, Ross M. Core stability and bicycling. *Curent Sports Medicine Reports.* 2010;9:155-60.

[39] Akuthota V, Ferreiro A, Moore T, Fredericson M. Core stability exercise principles. *Current Sports Medicine Reports.* 2008;7:39-44.

[40] Egaña M, Smith S, Green S. Revisiting the effect of posture on high-intensity constant-load cycling performance in men and women. *Euro J Appl Physiology.* 2007;99:495-501.

[41] Ashe MC, Scroop GC, Frisken PI, Amery CA, Wilkins MA, Khan KM. Body position affects performance in untrained cyclists. *Brit J Sports Med.* 2003;37:441-4.

In: Posture: Types, Exercises and Health Effects
Editor: Sarah A. Curran
ISBN: 978-1-63117-252-6

Chapter 8

Effects of Physical and Sporting Activities on Postural Stability in Children

Sonia Sahli,[1,2,*] PhD, Rym Baccouch,[2] MSc and Haithem Rebai,[3] PhD

[1]Research Unit Neurophysiology of Vigilance, Attention and Performance, Functional Explorations of the Nervous System Department, University Hospital Center Sahloul, Sousse University, Tunisia

[2]High Institute of Sport and Physical Education, Sfax University, Tunisia

[3]Research Unit Cardio Circulatory, Respiratory, Metabolic and Hormonal Adaptations to Exercise, Faculty of Medicine, Sousse University, Tunisia

Abstract

Concerns have been raised about the effects of physical and sporting activities on postural stability because it is an important determinant of both sports level and susceptibility to injuries. It may be also useful for clinicians interested in designing new rehabilitation therapies. While the sensorimotor control of posture and how physical and sport training affects it, have been widely studied in adults, surprisingly little is known about these effects in children's upright postural control. With this consideration, this chapter starts with a summary of the major results established in research undertaken on adults. It moves on to provide detailed analysis of the few available studies on the effects of physical activities on postural stability in children. In adults, several papers have shown that training may improve postural control by acting on the somatosensory, visual and vestibular systems involved in the control of a posture; or on the motor response through an increase of muscular strength. It has been observed that the nature of the movements involved in the practice of different sports would influence postural adaptation because training in a specific activity develops specific modalities of postural

[*] Corresponding author: sonia.sahli@yahoo.fr

control. Moreover, there is a relationship between competition level and postural control level. It is well known that postural control is not yet fully developed at birth. Indeed, it takes children almost a year to stand independently and many years thereafter to develop adult-like postural control. It has been suggested that postural control development in children is a result of how children integrate sensory information to achieve or maintain a desired posture. Recent studies showed that children before 12 years of age use sensory information different from adults to maintain an upright standing posture, and they do not compensate for sensory cue quality and magnitude changes as adults do. It was suggested that the practice of a physical exercise could modify the use of sensory information in children. To our best knowledge, there are 3 scientific works that investigate the relation between physical exercise training and postural stability in children. The performance of the postural control system improved in 5-6 year old children who practiced gymnastics and circus activities, whereas expertise in gymnastics did not seem to improve postural stability of 9-11 year old children. For older children aged 13 years of age, soccer training was efficient in improving the medial-lateral postural control. More studies are needed to gain a better understanding of how sensory organization develops in children and how physical and sporting activities affect their control of posture.

Keywords: Physical activities, sports, training, children, posture, postural stability

Introduction

Posture serves as a reference frame for the production of accurate movements that are vital for the activities of daily living like walking, stepping up the stairs, or standing in a bus. The most important function of it is to ensure the maintenance of equilibrium during the initiation and continuation of movement. [1] Beside appropriate strength and endurance capacity, the control of posture is considered to be an important indicator of musculoskeletal health and therefore could be of importance in view of clinical issues. [2] Postural stability refers to the inherent ability of a person to maintain, achieve or restore a specific state of balance under varied types of external disturbance to prevent falling while performing a range of activities from quite stance to complex athletic tasks. [3] Maintaining an upright standing posture results from an interaction between sensory information and motor action, [4] with visual, vestibular, and somatosensory systems as the main sources of sensory information. [5-9]

Several studies have shown that sport training may improve postural control by acting on these balance sensors [10-15] or on the motor response through an increase of muscular strength. [16] While the sensorimotor control of posture and how physical and sport training affects it, have been widely studied in adults, surprisingly little is known about these effects in children upright postural control. The nature of the development of postural control has intrigued researchers for many years.

A better understanding of how sensory organization develops in children and how physical and sporting activities affect their control of posture is important for many reasons. This knowledge would enable earlier detection of atypical postural development in children, provide better understanding and appreciation of the differences seen between individual and groups of children, and might also lead to improved interventions for children with pathological balance impairments.

Therefore, the aims of this chapter are:

(i) to provide a summary of the major results established in research undertaken on adults;
(ii) to determine if children integrate sensory information like adults to achieve or maintain a desired posture;
(iii) and to investigate the relation between physical exercise training and postural stability in children.

Physical and Sport Training Effects on Adult Postural Control

Training has widely been reported to be one of the best ways to improve balance. [10, 14, 15, 17-21] To understand how physical and sport training affects the control of posture in adults, several authors have attempted to assess differences in postural control between athletes and normal participants, [15, 22] athletes of different levels of sports expertise, [21-23] and athletes from different sports [15, 24-27] using only cross-sectional studies.

It seems that athletes, through years of sport experience, would have a different model of sensory integration, as compared to the non-athletes. [28] Previous studies have shown that training of each of the three levels of the sensorimotor chain may improve the maintenance of balance in complex conditions [i.e., with sensory deprivation] [29-31] indicating a positive effect of training on sensorimotor adaptability. [12-15] Differences in knee and ankle proprioception between trained athletes and non-athletes suggest that sport participation, by challenging sensorimotor systems, may improve balance. [13, 32] Some evidence in the literature suggested that superior balance among athletes is largely due to the repetitive training experiences that influence motor responses rather than to a greater sensitivity of the vestibular system. [33] Others argue that superior balance is the result of training experiences that influence a person's ability to attend to relevant proprioceptive and visual cues. [3] Although these experts may not agree on the mechanism, research suggested that changes in both sensory and motor systems influence balance performance. [26] Training experiences improving muscular strength of the lower limbs, [35] neuromuscular coordination, joint strength, and range of motion are also probable mechanisms that lead to improved balance. [13, 33, 36]

Compared to the non-athletes, athletes seem to have a special role of the visual vestibular interaction, when proprioception is altered. [28] They are coping more effectively than non-athletes with the multisensory integration challenges of the sway amplitude, while standing on an unstable support. [28] In addition, they seem to have a superior kinesthetic awareness, [37] increased safety margins [38] and an increased automaticity level. [39,40] Moreover, it has been suggested that compared to normal subjects, sportsmen adopted the reaction most appropriate to the task, using both past knowledge and short term memory very efficiently. [31, 41].

It has been asserted that the improvement of postural control seems to depend on the sport or the activity practiced. [15, 26, 42-45] The motor programs required are inevitably different from one sport to another. For example, individual sports like rhythmic gymnastics, dance, and artistic gymnastics require fine postural control especially in the one-legged stance. [24, 27, 46-49] In addition, many of gymnasts' skills require great strength and sometimes exaggerated joint range of motion. [50] Postural control performance of adult gymnasts has been widely investigated and compared with adult non-gymnasts using various physical exercises. [26, 24, 37, 47, 51, 52] In general, adult gymnasts present better performance of postural control than adult non-gymnasts, [24, 37, 47, 52] especially, during specific conditions such as the unipedal stance and on unstable bases of support. [24] However, Vuillerme et al. [24] showed that expert artistic gymnasts were not more stable than expert athletes in other sports while maintaining a bipedal and unipedal unperturbed stance suggesting that the eloquent sense of balance demonstrated by the gymnasts during their acrobatic movements does not provide any benefit to the achievement of more simple tasks. The good performances of judoists found in unusual situations, when compared to dancers and controls, could be due to the difference of balance modalities involved in judo and dance. [15] In fact, dancers train for long hours in a very stable environment and generate their own imbalance during their complex chained dynamic choreographic figures. Conversely, both in training and in competition, judoists are constantly subjected to unexpected movements imposed by their opponent in order to make them fall on the tatami. While ballet dancers often perform static postures, [15] modern dancers often perform dynamic movements [53] like other athletes; therefore the benefits of modern dance practice appear to be similar to those of athletics for improving balance. [54] A better balance control in static conditions has been observed in shooters compared to fencers. Shooting is a static activity based on a strict control of body posture where shooters are specifically trained to maintain a bipedal posture, as still as possible, in order to stabilize a weapon [28] or a rifle [55] in a stable environment; whereas, fencing is based on unceasingly moving situations of high-speed attacks forcing opponents to make constant spatial and temporal adjustments. [28]

Some team sports such as soccer or basketball demand a high degree of postural balance and quick movement changes. In fact, soccer players often perform single-leg reaching movements outside their base of support during passing, receiving, dribbling and shooting. [56] Basketball is particularly an activity of jumps, upper extremity passing, shooting, and dribbling skills requiring a great number of accelerations and decelerations, and direction changes. [57] Due to the training of these specific motor programs, the soccer players may have acquired balance ability much earlier than the controls [58] and made greater use of the somatosensory system during one-legged stance than basketball players and non-athletes. [27] It is expected that athletes participating in other land-based sports [such as volleyball and athletics], and who rarely support their body weight on one leg while playing, should have similar balance ability to basketball players. [27] However, when gymnasts were compared to basketball players, they develop less balance; whereas compared to soccer players they may have similar or better one-legged stance ability. [26, 27] Even though swimmers rarely use their anti-gravity muscles during training, no difference in stance stability between basketball players, swimmers, and non-athletes has been found. [27] This suggests that swimmers use their anti-gravity muscles to some extent when standing upright and walking in daily life, which may be sufficient to negate any negative influence of swimming training on their stance stability. [27] Highly skilled sportsmen were examined after long-term training with various kinds of sport and results showed that postural stability decreased in the following order: biathlon<judo<boat racing<water polo. [42]

Regarding the sensory organization of the standing balance, training in a specific activity develops specific postural adaptations [15, 28, 59] and lead to the discipline-oriented [37] optimal use of sensorimotor modalities responsible for body balance. [60] In this context, it has been asserted that postural stability in some sports such as in soccer, ballet dance, gymnastic, pistol shooting, cycling and surfing have a strong visual dependence, whereas, in judo, acrobatics, fencing, mountain bike training and triathlon have proprioceptive dependence. [15, 22, 43, 61-64] The vestibular input can be developed by Tai Chi, trampoline and some acrobatic activities. [65] It has been reported that team sport experts display improved perceptual abilities thanks to a relevant use of the visual information. [66] It has been documented that proprioceptive activities practice helps achieve better management of the conflict between visual and vestibular afferences, by improving particularly the proprioceptive afferences integration. [65] Perrin et al. [15] reported that judo and acrobatics training strengthens the accuracy of proprioceptive inputs better than dance that shifts with difficulty sensorimotor dominance from vision to proprioception. They also seem to be faster in switching between sensory inputs which suggests a better central integration. [14] Fencers and shooters exhibit more efficient balance control than the controls, both in eyes open and eyes closed conditions by developing particularly proprioceptive sensitivity. Because the role of vision is mainly attributed to the management of the action in both sports their vision cannot be used preferentially to control postural stability. [28] It has been reported that Tai Chi practitioners have better sensory organization ability as a result of using their vestibular inputs to mediate postural control than their non–Tai Chi counterparts. [67]

Due to the use of specific sensory cues to postural control, rhythmic gymnasts, as well as dancers and artistic gymnasts were more influenced by vision removing during an upright stance than adult non-gymnasts that practiced other physical exercises. [19, 46, 48, 52, 68] However, Roberston and Elliot [69] found that expert gymnasts are not influenced by vision. Vuillerme et al. [47] suggested that in challenging postures, gymnasts present the particularity of being less dependent on visual cues than other sportsmen for maintaining balance and explained this result by the fact that gymnasts can switch between visual and other sensory systems more efficiently and / or they have a more sensitive sensory system as compared to other sports experts. The expert soccer players developed the ability to shift the sensorimotor dominance from vision to proprioception [61] and built up a better internal representation of erect posture. [70] It has been argued that these specific postural adaptations or strategies of soccer players resulted from their experience in changing environment where vision is mainly used to collaborate with partner, to anticipate pass destination, and to fixate on peripheral aspects of the match, such as the positions and movements of other players. [23, 61, 71] Because ballet dancers use the visual afference as a major input to achieve a better balance regulation, [46] they are likely to do less well than controls in daily life situations where this afferent is missing. Conversely, judo training leads to the best performances in terms of maintaining a stable stance in all circumstances because they use proprioceptive afferences as an essential component of balance control. [72-75] These results confirm the hypothesis of a redistribution of postural control processing developed by judo practice [72-74] as in Tai Chi. [76]

A review of the literature [77] has provided support suggesting that there is a close relationship between the competition level and the postural control level in athletes. [18, 19, 23, 28, 37, 52, 55, 61, 70, 78] In general, a higher level of competition was associated with higher level of postural balance. This result was verified in soccer [61, 79] suggesting that professional soccer players may have improved their proprioceptive capacities to control the ball while maintaining balance [61] and would have a greater tone of the posterior leg muscles compared with the amateur soccer players. [80] The same results have been shown in golf [81], rifle shooting, [18] surfing [78] and combative sports. [72] It has been asserted that the more that the postural condition is specific for a certain sport and difficult, the greater is the difference in postural ability between expert sportsmen and non-expert sportsmen. [28, 52] Furthermore, the relationship between the postural ability level and the competition level should be stronger for the sport activities that involve maintaining balance with unstable support. [58]

The better postural abilities in the national and international levels than in the local level in surfing have been linked to better sensorimotor adaptability to training or better genetic disposition [78] or to the greater number of practice sessions and of hours spent in training that may have facilitated the improved postural adaptations. It seems that intensive sports training involves learning automatic movements that become stereotyped and reproducible. [57]

It appears that expert sportsmen use predominantly certain sensorial information to regulate their posture according to the requirement of their discipline. [14, 47] The relationship between the contribution of vision in postural control and the expertise level is specific to disciplines characterized by a strong sensorimotor dominance of vision. [15, 82] For example, in expert dancers [19] the visual contribution to postural maintenance increased as the level of competition increases; whereas with soccer players [61] and surfing, [78] the participation of visual cues in postural regulation decreases as the level of competition increases suggesting that the lower dependence on vision for postural control is attributed to the acquisition of this ability to control the ball without watching it in professional soccer players [61] and to a greater competence to perform technical figures while maintaining balance in national and international surfers. [78] Chapman et al. [22] have previously suggested that expert surfers could shift the sensorimotor dominance from vision to proprioception for postural maintenance. A better efficiency of proprioceptive inputs in high level sportsmen than in lower level sportsmen have been reported and attributed either to a higher sensitivity of sensory receptors [15] and / or a better knowledge of the orientation of body axis and verticality. [83]

Postural Control Development in Children

Significant concerns about the development of balance control in humans during the life have been raised. [84-87] Standing upright and maintaining balance is one of the major milestones that characterize the motor development of early childhood. To control a posture requires an accurate and reliable relationship between sensation and action which is not yet fully developed at birth. Indeed, it takes infants almost a year to stand independently and many years thereafter to develop adult-like postural control. [88]

Although, several studies asserted that young children exhibit a greater magnitude of postural sway than adults during a quiet standing position [89-93] and that postural control improvement during childhood is characterized by a decreasing magnitude [94, 95] and frequency [96] of postural sway with a non-linear rate, [97] there is a controversy concerning when adult-like begin to appear. It has been reported that it is around the age of 7–8 years that adult-like balance control strategies and the ability to resolve sensory conflict begin to appear. [85, 87] Based on data from large sample sizes and specific age ranges, many studies demonstrated that mature responses do not appear until much later in childhood or adolescence, whether considering postural sway with quiet stance or when evaluating the use of sensory information. Rival et al. [98] found that children until 10-years-old are less efficient than adults in the control of either static or dynamic balance. Similarly, Wu et al. [99] conclude that it takes more than 10 years for children to refine the variability partitioning strategy and achieve an adult-like capability. Findings of other studies showed that children aged 9–12 years of age presented postural control performance similar to adults, [100] and achieve adult-like sensory integration during stance at 12-years-old. [89] Peterka and Black [101] found that children younger than 15 years old demonstrated increased postural sway compared to adults when all sensory information was available and accurate; Hirabayashi and Iwasaki [102] proposed that generalized postural stability had not reached adult level by age 15 years, nor had vestibular function for resolving sensory conflict. Garcia et al. [103] concluded that adult-like performance in children's postural control is task-dependent and might vary as a function of sensory manipulation.

The logical explanation for the postural stability differences between children and adults seems to be the immaturity of the sensory systems. From a physiological viewpoint, the visual [104, 105] and vestibular [106, 107] systems become mature well before balance performance is adult-like. Therefore, differences in postural control between children and adults seem to be a result of how children integrate visual [94, 108], vestibular [109] and somatosensory [110] information into motor action to achieve or maintain a desired posture. [89]

Some studies have shown that visual information affects [90, 98] postural control of children while others have not. [93, 100, 111] Whatever the result considered, the integration of visual information seems to improve during childhood. From age 4 to 6 years, the visual information is considered as the dominant sensory information for postural control [84] where children use a ballistic strategy (open-loop control) with large and fast corrections which may be accounted for by the predominance of feedback responses. [110]

Critical changes around the 7–8 years of age occur when children seem to have some difficulties in minimizing the magnitude of the centre of pressure (COP) displacements. Whereas the amount of activity required decreased as result of the increase of the range and the decrease of the speed. [98, 108] These critical changes in postural control characteristics have been reported earlier in childhood (4–6 years) in other studies. [112] After the critical stage, a progress from a ballistic strategy to an integrated open-loop and closed-loop mode of control occurs resulting in reduced both the range and the speed of COP displacements. [98] This pattern has been interpreted as a change from a speed to an accuracy strategy, and then to a combination of these two constraints; [85] and as a change from a reactive strategy [feedback responses] to an effective feedforward postural adjustments. [110] Then, through upright posture experience in standing and walking, infants may learn to refine the sensorimotor relationship and thus to better incorporate sensory information in the postural control system. [89] Moreover, it has been not only shown that the integration of sensory information in children before 12 years old is different from adults, [89] but also that these children do not compensate for sensory cue quality [113] and magnitude changes [114] as adults do. For vestibular perception several studies reported that vestibular maturity for resolving sensory conflict occurs at a later age of 15–16 years. [89, 101, 115, 116] In further efforts to explain the nature of postural control development, the influence of anthropometric characteristics has been investigated; and results of these investigations vary from no correlation between sway parameters and anthropometric characteristics when children were asked to maintain static stance with or without visual feedback [117] to approximately 20% of contribution of these characteristics to the variability in equilibrium scores. [89, 95]

According to Assaiante and Amblard [84] postural control development in children is linked to the interaction of mechanical and sensory development: the first step for children consists of building a repertoire of postural strategies; the second step consists of learning to select the most appropriate postural strategy depending on the ability to anticipate on the consequence of the movement in order to maintain balance control and the efficiency of the task. [1]

Another explanation of postural control improvement during childhood has been presented: a gradual refinement of both the localization and the level of muscle activity as electromyographic activity involved in static postural task performance decreased with increasing age. [118] Furthermore, as leg muscles contract in an effort to control a stable upright posture, proper muscular strength and architecture might be beneficial for postural control. [119] Thus, the poor postural control in young children could be explained by the lack of the skeletal musculature development. [120] In this context, it has been reported that deficits in postural control and muscle strength have been observed in children compared with that in young healthy adults. [121, 122] Conversely, no significant associations were found between variables of postural control and muscle strength in children. [123]

Physical and Sport Training Effects on Postural Stability in Children

Peterson et al. [89] suggested that age alone accounted for only 16% of the total postural stability variance in children, with anthropometric characteristics and gender adding another 4% contribution. Stambolieva et al., [28] believe that a fair part of the remaining 80% comes from physical activity. On the other hand, it has been asserted that postural control development requires not only the maturation of the nervous and musculoskeletal systems, but also the coordination between these systems and the experience of practicing postural tasks. [124] If these developmental changes are related to experience, then specific physical and sporting activities training could provide a support to develop postural control. Unfortunately, few studies on the effects of physical and sporting activities on postural stability in healthy children are available and surprisingly only 3 activities have been studied: gymnastics, circus activities and soccer.

The influence of gymnastics training on postural control have been investigated in children aged 5–7 and 9–11 years-old with and without the use of visual information. Results showed that younger gymnasts were more stable than younger nongymnasts indicating that regular and systematic training of physical exercise, such as gymnastics, might improve postural control approximately during the fifth year of life. Such improvements have been related to the use of sensory cues that children have available to maintain a desired posture.[103] Moreover, the controversial observations regarding the visual information effects on postural control reported in children could be attributed to some uncontrolled factors that may significantly affect postural control such as regular physical training. It is interesting to observe that training effects for the younger children seem to occur even with relatively short period (an average of 1.65 hour) of gymnastics experience. [103]

In the case of older children, it seems that even long periods (an average of 3.95 hour) of gymnastics experience does not affect postural performance. [103] In fact, the benefits of gymnastics training on postural control of the older gymnasts (9–11 year-olds) were not observed suggesting that older gymnasts do not gain any advantage from their specific training or that maintaining a bipedal upright stance may not challenge the postural control system enough in order to discriminate the performance of postural control between gymnasts and non-gymnasts, especially at this age. [103] Therefore, it has been concluded that postural control development in older children may be age-related and also task-dependent, [89, 103] similar to adults. [125] Postural control of children under different stances should be investigated in future studies to confirm this conclusion.

The effect of circus activity training on postural control has been investigated in children aged between 5 and 6 years, critical stage of postural development according to Woollacott et al. [112] Postural control results showed that circus group children had better performances than control group children suggesting that training in circus activities seems to have an effect on children postural control. In fact, performing acrobatics and dynamic equilibrium tasks in challenging conditions similar to those involved in circus activities, places great demand on the postural control system. Even if their sport consists essentially to control body movement, they are also trained to stand still for a certain period in various postures. The better postural performance demonstrated in circus group has been explained by a better ability to use proprioceptive and somesthesic information. [126]

By showing a better postural control for both the unspecific (static balance condition) and specific (dynamic condition) postures in circus group children, these results argue in favor of the general motor ability hypothesis, [127] predicting that any human skill should be observed among various test conditions [128] and indicate that circus training can develop sensorimotor adaptabilities transferable to general postural control. [126] Similarly, adult elite gymnasts have been shown to have better dynamic and static balance than novices. [129, 130] However, in support of the Henry's specificity hypothesis stipulating the idea that improvement of performance due to learning is specific to the task and not directly transferable, [31] the adult ballet dancers have been shown to perform the bipedal posture similarly to controls; but when performing a posture consistent with their training, the unipedal one, they had significantly better postural performances suggesting that these athletes develop specific modalities of balance that are not transferable to posture control in daily activities. [132]

Results of posture's difficulty classification showed a main effect of postural condition on postural control of both groups indicating that balance control depends on the difficulty of the task being performed and that circus group children could not counteract the difficulty induced by the task's characteristics similarly to controls. [126] The difficulty in maintaining the dynamic conditions is essentially due to biomechanical constraints and sensorial characteristics. [133] It has been previously shown that postural sway increased similarly for adult expert gymnasts and adult experts in other non-gymnastics sports as the posture difficulty increased. [24, 128] It would be possible that the control group balance improved through the play activities and games they engaged in over the previous 2 years and transferred to their performance of the involved postures. [126]

Removal of vision decreased the postural control of both groups of children in all postures [126] which is in agreement with previous studies investigating adult postural balance. [19, 61, 82] However, studies analyzing children postural balance provided conflicting evidence about the role of vision in the control of posture. These conflicting findings are probably due to the fact that these studies used different dependent variables in different motion planes with a variety of task conditions over a wide age ranges. [108] Furthermore, removal of vision affected circus group children and controls similarly suggesting that circus training could not nullify the common effect of vision on postural control of children [126] as reported for adult gymnasts [52, 68] and skiers. [21] In contrast, gymnasts were found to be less dependent on visual cues than untrained subjects. [130, 134] Moreover, it has been suggested that postural control of a circus group of children and controls became progressively altered by the removal of vision as the difficulty of the posture increased. [126] In accordance with previous adult studies, [24, 135] this suggested that the role of vision increased when the difficulty of the stance increased. [126]

The influence of soccer training has also been studied; but for older children (pre-adolescents) aged 13 years-old. [58] Results of this study showed that: (1) soccer training was efficient in improving the medial-lateral postural control in young boys; (2) athletes developed specific postural strategies characterized by decreased COP frequency and lower reliance on vision.

Postural differences between soccer players and non-athletes in the medial-lateral plane probably indicates that playing soccer and the specific exercises improve medial-lateral stability more than anterior-posterior stability. It is also possible that postural control in children matures faster in the anterior-posterior than in the medial-lateral plane. This suggests that when a need for stability improvement is considered, priority be given to exercises which reinforce the medial-lateral stability. [58]

To identify possible neurophysiologic changes or sensorimotor modalities which might have emerged as a result of soccer, several explanation have been advanced. The lower sway frequency that accompanied the better postural stability in preadolescent soccer players is a good indicator of postural strategies development due to maturation and / or training. [58] As explained above, very young children have been shown to use higher rate of postural corrections [111] which may be accounted for by the predominance of feedback responses. The effective feedforward postural adjustments turn up later and partially replace this reactive strategy which results in a decrease of the COP frequency. [110] Due to the specific training, the soccer players may have acquired this ability much earlier than the controls. [58] In addition, experience in dynamic standing tasks improves postural stability by increasing the safety margins [38] which allow for smoother sway excursions probably resulting in slowing down the rate of postural corrections. [58] Increased automaticity level due to learning of new postural activity may also leads to better performance. [39, 40] Thus, the decreased COP frequency in soccer players may be accounted for by eliminating some ‘‘noise’’ [136] which was introduced by redundant motor units involved in postural regulation at the early stage of motor learning. [58]

Nevertheless, it is possible that there are many other factors related to physical activity such as its type, intensity, frequency, and timing that may significantly modify and optimize the children postural control long before their biological maturation. [58] Therefore, further investigation of these factors effects on postural stability development in children is needed as it may offer new inspirations in physiotherapy and in sports requiring special balance abilities. Moreover, the beneficial effects of these three activities (gymnastics, circus and soccer) on postural stability of children have only been demonstrated by simple comparisons with sedentary subjects. It has been asserted in adults that the improvement of postural control due to sports training seems to depend on the sport or the activity practiced. [15, 26, 42-45] Thus, to better understand how physical and sport training affects the control of posture in children, further comparisons between younger athletes from different sports should be conducted as they may offer a support for determining which of these sports appeared to mostly improve balance control of children in unexpected situations. It may be also inspiring for clinicians interested in designing new rehabilitation therapies for injured children or nonactive children with balance instability.

Conclusion

Training has widely been reported to be one of the best ways to improve balance as it has a positive effect on sensorimotor adaptability. Compared to the non-athletes, athletes seem to have a special role of the visual vestibular interaction, when proprioception is altered, increased safety margins, improve muscular strength of the lower limbs and a better neuromuscular coordination. The improvement of postural control due to training seems to depend on the sport or the activity practiced.

There is a close relationship between the competition level and the postural control level in athletes. In general, a higher level of competition has been associated with higher level of postural balance and linked either to better sensorimotor adaptability to training or better genetic disposition or to the intensive sports training involving learning automatic movements.

Significant concerns about the development of balance control in humans during life have been raised. Improvement of postural control during childhood has been characterized by a decreasing magnitude and frequency of postural sway and a non-linear rate of decrease. Differences in postural control between children and adults seem to be a result of how children integrate sensory information to achieve or maintain a desired posture. There is controversy concerning when adult-like begin to appear. Some studies reported that it is around the age of 7–8 years of age that adult-like balance control strategies become apparent, whereas, others demonstrated that mature responses do not appear until much later in childhood or adolescence.

Through upright posture experience in standing and walking, children may learn to better incorporate sensory information in the postural control system; and to compensate for sensory cue quality and magnitude changes as adults do. Postural control development in children has also been linked to the interaction of mechanical and sensory development or to a gradual refinement of both the localization and level of muscle activity, or to anthropometric characteristics.

Few studies on the effects of physical and sporting activities on postural stability in healthy children are available. Results indicated that regular and systematic training of physical exercise, such as gymnastics, might improve postural control approximately during the fifth year of life. Whereas, older gymnasts do not gain any advantage from their specific training to better perform a bipedal upright stance. Circus activity training at a rate of twice a week improves postural control of children aged between 5 and 6 years old. As the difficulty of the postural task increases, children having this experience in circus training do not control their balance better than controls. Circus training could not nullify the common effect of removing vision on postural control. Moreover, this effect increased when the difficulty of the postural task increased. Soccer training was found to be efficient in improving the medial-lateral postural control in young boys aged 13-years-old. Pre-adolescent soccer players also seem to develop specific postural strategies characterized by decreased COP frequency and lower reliance on vision.

To better understand how physical and sport training affects the control of posture in children, further studies investigating the effects of many other factors related to the physical or sports activity training such as type, intensity, frequency, and timing are needed as they may offer a support for determining which sports activity could be incorporated into rehabilitative programs for injured children or nonactive children with balance instability.

References

[1] Assaiante C, Mallau S, Viel S, Jover M, Schmitz C. Development of postural control in healthy children: a functional approach. *Neural Plast.* 2005;12:109-18.

[2] Geldhof E, Cardon G, De Bourdeaudhuij I, Danneels L, Coorevits P, Vanderstraeten G et al. Static and dynamic standing balance: test-retest reliability and reference values in 9 to 10 year old children. *Eur J Pediatr.* 2006;65:779-86.

[3] Pollock AS, Durward BR, Rowe PJ, Paul JP. What is balance? *Clin Rehabil.* 2000;14:402-06.

[4] Horak FB, Macpherson JM. Postural orientation and equilibrium. In: Rowell LB, Shepard JT (eds). Handbook of Physiology, Oxford University Press, New York; 1996. p. 255-92.

[5] Nashner LM. Analysis of stance posture in humans. In: Towe AL, Luschei ES (eds). Motor coordination. Handbook of behavioral neurology. Vol 5. New York: Plenum Press NY; 1981. p. 527-565.

[6] Horak FB, Shupert CL, Dietz V, Horstmann G. Vestibular and somatosensory contributions to responses to head and body displacements in stance. *Exp Brain Res.* 1994;100:93-106.

[7] Massion J. Postural control system. *Curr Opin Neurobiol.* 1994;4:877–87.

[8] Winter D. Human balance and posture control during standing and walking. *Gait Posture.* 2000;3:193-214.

[9] Vuillerme N, Pinsault N, Vaillant J. Postural control during quiet standing following cervical muscular fatigue: effects of changes in sensory inputs. *Neurosci Lett.* 2005;378:135-9.

[10] Hu MH, Woollacott MH. Multi-sensory training of standing balance in older adults: kinetic and electromyographic postural responses. *J Gerontol.* 1994;9:69-71.

[11] Perrin Ph, Gauchard GC, Perrot C, Jeandel C. Effects of physical and sporting activities on balance control in elderly people. *Br J Sports Med.* 1999;33:121-26.

[12] Crotts D, Thompson B, Nahom M, Ryan S, Newton RA. Balance abilities of professional dancers on select balance tests. *JOSPT*. 1996;23:12-7.

[13] Lephart SM, Giraldo JL, Borsa PA, Fu FH. Knee joint proprioception: a comparison between female intercollegiate gymnasts and controls. *Knee Surg Sports Traumatol Arthrosc.* 1996;4:121-4.

[14] Perrin Ph, Schneider D, Deviterne D, Perrot C, Constantinescu L. Training improves the adaptation to changing visual conditions in maintaining human postural control in a test of sinusoidal oscillation of the support. *Neurosci Lett.* 1998;245:155-8.

[15] Perrin P, Deviterne D, Hugel F, Perrot C. Judo, better than dance, develops sensorimotor adaptabilities involved in balance control. *Gait Posture*. 2002;15:187-94.

[16] Wolfson L, Whipple R, Derby C, Judge J, King M, Amerman P, et al. Balance and strength training in older adults: intervention gains and Tai Chi maintenance. *J Am Geriatr Soc*. 1996;44:498-506.

[17] Campbell AJ, Robertson MC, Gardner MM, Norton RN, Tilyard MW, Buchner DM.Randomised controlled trial of a general practice programme of home based exercise to prevent falls in elderly women. *Br Med J.* 1997;315:1065-9.

[18] Era P, Konttinen N, Mehto P, Saarelas P, Lyytinen H. Postural stability and skilled performance: a study on top-level and naïve rifle shooters. *J Biomech.* 1996;29:301-6.

[19] Golomer E, Cremieux J, Dupui P, Isableu B, Ohlmann T. Visual contribution to self-induced body sway frequencies and visual perception of male professional dancers. *Neurosci Lett.* 1999;267:189-92.

[20] Bringoux L, Marin V, Nougier V, Barraud PA, Raphel C. Effects of gymnastics expertise on the perception of body orientation in the pitch dimension. *J Vestib Res.* 2000;10:251-8.

[21] Noé F, Paillard T. Is postural control affected by expertise in alpine skiing? *Br J Sports Med.* 2005;39:835-7.

[22] Chapman DW, Needham KJ, Allison GT, Lay B, Edwards DJ. Effects of experience in dynamic environment on postural control. *Br J Sports Med.* 2008;42:16-21.

[23] Paillard T, Noé F. Effect of expertise and visual contribution on postural control in soccer. *Scand J Med Sci Sports.* 2006;16:345-8.

[24] Vuillerme N, Danion F, Marin L, Boyadjian A, Prieur JM, Weise I et al. The effect of expertise in gymnastics on postural control. *Neurosci Lett.* 2001;303:83-6.

[25] Vuillerme N, Nougier V. Attentional demand for regulating postural sway: the effect of expertise in gymnastics. *Brain Res Bull.* 2004;63:161-5.

[26] Bressel E, Yonker JC, Kras J, Heath EM. Comparison of static and dynamic balance in female collegiate soccer, basketball, and gymnastics athletes. *J Athl Train.* 2007;42: 42-46.

[27] Matsuda S, Demura S, Uchiyama M. Centre of pressure sway characteristics during static one-legged stance of athletes from different sports. *J Sports Sci.* 2008;26:775-9.

[28] Stambolieva K, Diafas V, Bachev V, Christova L, Gatev P. Postural stability of canoeing and kayaking young male athletes during quiet stance. *Eur J Appl Physiol.* 2012;112:1807-15.

[29] Hlavacka F, Mergner T, Schweigart G. Intervention of vestibular and proprioceptive inputs for human selfmotion perception. *Neurosci Lett.* 1992;138:161-164.

[30] Perrin Ph, Béné MC, Perrin C, Durupt D. Ankle trauma significantly impairs posture control: a study in basketball players and controls. *Int J Sports Med.* 1997;18:387-92.

[31] Vidal PP, Berthoz A, Millavoye M. Difference between eye closure and visual stabilization in the control of posture in man. *Aviat Space Environ Med.* 1982;53: 166-78.

[32] Aydin T, Yildiz Y, Yildiz C, Atesalp S, Kalyon TA. Proprioception of the ankle: a comparison between female teenaged gymnasts and controls. *Foot Ankle Int.* 2002;23:123-9.

[33] Balter SGT, Stokroos RJ, Akkermans E, Kingma H. Habituation to galvanic vestibular stimulation for analysis of postural control abilities in gymnasts. *Neurosci Lett.* 2004;366:71-5.

[34] Ashton-Miller JA, Wojtys EM, Huston LJ, Fry-Welch D. Can proprioception really be improved by exercises? *Knee Surg Sports Traumatol Arthrosc.* 2001;9:128-136.

[35] Gauchard GC, Jeandel C, Tessier A, Perrin PP. Beneficial effect of proprioceptive physical activities on balance control in elderly human subjects. *Neurosci Lett.* 1999;273:81-4.

[36] Paterno MV, Myer GD, Ford KR, Hewett TE. Neuromuscular training improves single-limb stability in young female athletes. *JOPST.* 2004;34:305-16.

[37] Davlin CD. Dynamic balance in high level athletes. *Percept Mot Skills.* 2004;98: 1171-6.

[38] Patton JL, Lee WA, Pai YC. Relative stability improves with experience in a dynamic standing task. Exp Brain Res. 2000;135:117-26.

[39] Logan GD. Automaticity, resources, and memory: theoretical controversies and practical implications. *Hum Factors.*1988;30:583-98.

[40] Milton JG, Small SS, Solodkin A. On the road to automatic: dynamics aspects in the development of expertise. *J Clin Neurophysiol.* 2004;21:134-43.

[41] Thomson JA. Is continuous visual monitoring necessary in visually guided locomotion? *J Exp. Psychol Hum Percept Perform.* 1983;9:427-43.

[42] Arkov VV, Abramova TF, Nikitina TM, Ivanov VV, Suprun DV, Shkurnikov MU et al. Comparative study of stabilometric parameters in sportsmen of various disciplines. *Bulletin of Experimental Biology Medicine.* 2009;147:233-35.

[43] Nagy E, Toth K, Janositz G, Kovacs G, Feher-Kiss A, Angyan L et al. Postural control in athletes participating in an ironman triathlon. *Eur J Appl Physiol.* 2004;92:407-13.

[44] Gerbino P, Griffin E, Zurakowski D. Comparison of standing balance between female collegiate dancers and soccer players. *Gait Posture.* 2007;26:501-7.

[45] Ford KR, Hewett TE, Paterno MV, Myer GD. Neuromuscular training improves single-limb stability in young female athletes. *JOSPT.* 2004;34:305-16.

[46] Hugel F, Cadopi M, Kohler F, Perrin Ph. Postural control of ballet dancers: a specific use of visual input for artistic purposes. *Int J Sports Med.* 1999;20:86-92.

[47] Vuillerme N, Teasdale N, Nougier V. The effect of expertise in gymnastics on proprioceptive sensory integration in human subjects. *Neurosci Lett.* 2001;311:73-6.

[48] Calavalle AR, Sisti D, Rocchi MBL, Panebianco R, Del Sal M, Stocchi V. Postural trials: expertise in rythmic gymanastics increases control in lateral directions. *Eur J Appl Physiol.* 2008;104:643-9.

[49] Lamoth CJC, Van Lummel RC, Beek PJ. Athletic skill level is reflected in body sway: a test case for accelometry in combination with stochastic dynamics. *Gait Posture.* 2009;29:546-51.

[50] Hay JG. The Biomechanics of Sports Techniques. 4th ed. Englewood Cliffs, NJ: Prentice Hall; 1993. p. 528.
[51] Gautier G, Thouvarecq R, Vuillerme N. Postural control and perceptive configuration: influence of expertise in gymnastics. *Gait Posture.* 2008;28:46-51.
[52] Asseman F, Caron O, Crémieux J. Are there specific conditions for which expertise in gymnastics could have an effect on postural control and performance? *Gait Posture.* 2008;27:76-81.
[53] Sides SN, Ambegaonkar JP, Caswell SV. High incidence of shoulder injuries in collegiate modern dance students. *Athletic Therapy Today.* 2009;14:43-6.
[54] Ambegaonkar JP, Caswell SV, Winchester JB, Shimokochi Y, Cortes N, Caswell AM. Balance comparisons between female dancers and active nondancers. *Research Quarterly Exercise Sport.* 2013;84:24-9.
[55] Aalto H, Pyykkö I, Ilmarinen R, Kähkönen E, Starck J. Postural stability in shooters. *ORL J Otorhinolaryngol Relat Spec.* 1990;52:232-8.
[56] Golomer E, Vandewalle H, Lefevre P, Pérès G. Equilibre et pied d'appui du footballeur. In: Doin, editor. Les troubles de l'équilibre. Paris; 1992. p. 137-41.
[57] Leroy D, Polin D, Tourny-Chollet C, Weber J.Spatial and temporal gait variable differences between basketball, swimming and soccer players. *Int J Sports Med.* 2000;21:158-62.
[58] Biec E, Kuczynski M. Postural control in 13-year-old soccer players. *Eur J Appl Physiol.* 2010;110:703-8.
[59] Asseman F, Caron O, Crémieux J. Is there a transfer of postural ability from specific to unspecific postures in elite gymnasts? *Neurosci Lett.* 2004;358:83-6.
[60] Rose DJ. Motor control and learning. Boston: Allyn & Bacon;1997.
[61] Paillard T, Noé F. Effect of expertise and visual contribution on postural control in soccer. *Scand J Med Sci Sports.* 2006;16:345-8.
[62] Schmit JM, Regis DI, Riley MA. Dynamic patterns of postural sway in ballet dancers and track athletes. Exp Brain Res. 2005;163:370-8.
[63] Lion A, Gauchard GC, Deviterne D, Perrin PhP. Differentiated influence of off-road and on-road cycling practice on balance control and the related-neurosensory organization. *J Electromyogr Kinesiol.* 2009;19:623-30.
[64] Herpin G, Gauchard GC, Lion A, Collet P, Keller D, Perrin PP. Sensorimotor specificities in balance control of expert fencers and pistol shooters. *J Electromyogr Kinesiol.* 2010;20:162-9.
[65] Caillet G, Bosser G, Gauchard GC, Chau N, Benamghar L, Perrin PP. Effect of sporting activity practice on susceptibility to motion sickness. *Brain Res Bull.* 2006;69:288-93.
[66] Kioumourtzoglou E, Derri V, Tzetzis G, Theodorakis Y. Cognitive, perceptual, and motor abilities in skilled basketball performance. *Percept Mot Skills.* 1998;86:771-86.
[67] Fong SM, Ng GY. The effects on sensorimotor performance and balance with Tai Chi training. *Arch Phys Med Rehabil.* 2006;87:82-7.
[68] Bardy BG, Laurent M. How is body orientation controlled during somersaulting? *J Exp Psychol Hum Percept Perform.* 1998;24:963-77.
[69] Robertson S, Elliott D. Specificity of learning and dynamic balance. *Res Q Exerc Sport.* 1996;67:69-75.

[70] Paillard T, Montoya R, Dupui P. Specific postural adaptations according to the throwing techniques practiced in competition level judoists. *J Electromyogr Kinesiol.* 2007;17:241-4.

[71] Williams AM, Davids K, Burwitz L, Williams JG. Visual search strategies in experienced and inexperienced soccer players. *Res Q Exerc Sport.* 1994;65:127-235.

[72] Perrot C, Deviterne D, Perrin Ph. Influence of training on postural and motor control in a combative sport. *J Hum Mov Studies.* 1998;35:119-36.

[73] Mesure S, Crémieux J. The effect of judo training on postural control assessed by accelerometry. In: Brandt Th, editor. Proceeding of the Xth international symposium on disorders of posture and gait. New York: Stuttgart; 1992. p. 302-6.

[74] Barrault D, Brondani JC, Rousseau D. Médecine du Judo. Paris:Masson; 1991.

[75] Perrot C, Mur JM, Mainard D, Barrault D, Perrin PP. Influence of trauma induced by judo practice on postural control. *Scand J Med Sci Sports.* 2000;10:292-7.

[76] Hain TC, Fuller L, Weil L, Kotsias J. Effects of Tai Chi on balance. *Arch Otolaryngol Head Neck Surg.* 1999;125:1191-5.

[77] Hrysomallis C. Balance ability and athletic performance. *Sports Med.* 2011;41:221-232.

[78] Paillard T, Margnes E, PortetM. Arnaud Breucq A. Postural ability reflects the athletic skill level of surfers. *Eur J Appl Physiol.* 2011;111:1619-23.

[79] Butler RJ, Southers C, Gorman PP, Kiesel KB, Plisky PJ. Differences in soccer players' dynamic balance across levels of competition. *J Athl Train.* 2012;47:616-20.

[80] Gagey PM, Weber B. Posturologie. Paris: Masson, 1999.

[81] Sell TC, Tsai YS, Smoliga JM, Myers JB, Lephart SM. Strength, flexibility, and balance characteristics of highly proficient golfers. *J Strength Cond Res.* 2007;21: 1166-71.

[82] Paillard T, Costes-Salon MC, Lafont C, Dupui P. Are there differences in postural regulation according to the level of competition in judoists? *Br J Sports Med* 2002;36:304-5.

[83] Paillard T, Bizid R, Dupui P. Do sensorial manipulations affect subjects differently depending on their postural abilities? *Br J Sports Med.* 2007;41:435-8.

[84] Assaiante C, Amblard B, An ontogenetic model for the sensorimotor organization of balance control in humans. *Human Mov. Sci.* 1995;14:13-43.

[85] Kirshenbaum N, Riach CL, Starkes JL. Non-linear development of postural control and strategy use in young children: a longitudinal study. *Exp Brain Res.* 2001;140:420-431.

[86] Nougier V, Bard C, Fleury M, Teasdale N. Contribution of central and peripheral vision to the regulation of stance: developmental aspects. *J Exp Child Psychol.* 1998; 68:202-215.

[87] Shumway-Cook A, Woollacott M. The growth of stability: postural control from a developmental perspective. *J Mot Behav.* 1985;17:131-47.

[88] Chen LC, Metcalfe JS, Chang TY, Jeka JJ, Clark Je. The development of infant upright posture: sway less or sway differently? *Exp Brain Res.* 2008;186:293-303.

[89] Peterson ML, Christou E, Rosengren KS. Children achieve adult-like sensory integration during stance at 12-years-old. *Gait Posture.* 2006;23:455-63.

[90] Riach C, Starkes J. Velocity of centre of pressure excursions as an indicator of postural control systems in children. *Gait Posture.* 1994;2:167-72.

[91] Riach C, Starkes J. Visual fixation and postural sway in children. *J Mot Behav.* 1989;21:265-76.

[92] Usui N, Maekawa K, Hirasawa Y. Development of the upright postural sway of children. *Dev Med Child Neurol.* 1995;37:985-96.
[93] Zernicke RF, Gregor RJ, Cratty BJ. Balance and visual proprioception in children. J *Hum Mov Stud.* 1982;8:1-13.
[94] Hatzitaki V, Zisi V, Kollias I, Kioumourtzoglou E. Perceptual motor contributions to static and dynamic balance control in children. *J Mot Behav.* 2002; 34:161-170.
[95] Odendrick P, Sandstedt P. Development of postural sway in the normal child. *Human Neurobiol.* 1984;3:241-4.
[96] Figura F, Cama G, Capranica L, Guidetti L, Pulejo C. Assessment of static balance in children. *J Sports Med Phys Fitness.* 1991;31:235-42.
[97] Forssberg H, Nashner LM. Ontogenetic development of postural control in man: adaptation to altered support and visual conditions during stance. *J Neurosci.* 1982;2:545-52.
[98] Rival C, Ceyte H, Olivier I. Developmental changes of static standing balance in children. *Neuroscience Letters.* 2005;376:133-6.
[99] Wu J, McKay S, Angulo-Barroso R. Center of mass control and multi-segment coordination in children during quiet stance. *Exp Brain Res.* 2009;196:329-39.
[100] Taguchi K, Tada C. Change in body sway with growth of children. In: Amblard B, Berthoz A, Clarac F, editors. Posture and gait: development, adaptation, and modulation. Amsterdam: Elsevier; 1988. p. 59-65.
[101] Peterka RJ, Black FO. Age-related changes in human posture control: sensory organization tests. *J Vest Res.* 1990;1:73-85.
[102] 1023. Hirabayashi S, Iwasaki Y. Developmental perspective of sensory organization on postural control. *Brain Dev.* 1995;17:111-3.
[103] Garcia C, Barela JA, Viana AR, Barela AM. Influence of gymnastics training on the development of postural control. *Neurosci Lett.* 2011;492:29-32.
[104] Neuringer M, Jeffrey BG. Visual development: neural basis and new assessment methods. *J Pediatr.* 2003;87-95.
[105] Brecelj J. From immature to mature pattern ERG and VEP. *Doc Opthalmol.* 2003;107:215-24.
[106] Dayal VS, Farkashidy J, Kokshanian A. Embryology of the ear. *Can J Otolaryngol.* 1973;2:136-42.
[107] Bergstrom B. Morphology of the vestibular nerve. Part I. Anatomical studies of the vestibular nerve in man. *Acta Otolaryngol.* 1973;76:162-72.
[108] Riach C, Starkes J. Stability limits of quiet standing postural control in children and adults. *Gait Posture.* 1993;1:105-11.
[109] Perrin P, Perrin C. Sensory afferences and motor control of equilibrium using static and dynamic posture tests. *Ann Otolaryngol Chir Cervicofac.* 1996;113:133-46.
[110] Haas G, Diener HC, Rapp H, Dichgans J. Development of feedback and feedforward control of upright stance. *Dev Med Child Neurol.* 1989;31:481-8.
[111] Riach CL, Hayes KC. Maturation of postural sway in young children. *Dev Med Child Neurol.* 1987;29:650-658.
[112] Woollacott M, Debu B, Mowatt M. Neuromuscular control of posture in the infant and child: is vision dominant? *J Mot Behav.* 1987;19:167-86.
[113] Godoi D, Barela JA. Body sway and sensory motor coupling adaptation in children: effects of distance manipulation. *Dev Psychobiol.* 2008;50:77-87.

[114] Rinaldi NM, Polastri PF, Barela JA. Age-related changes in postural control sensory reweighting. *Neurosci Lett.* 2009;467:225-229.
[115] Rine RM, Rubish K, Feeney C. Measurement of sensory system effectiveness and maturational changes in postural control in young children. *Pediatr Phys Ther.* 1998;10:16-22.
[116] Steindl R, Kunz K, Schrott-Fischer A, Scholtz AW. Effect of age and sex on maturation of sensory systems and balance control. Dev Med Child Neurol. 2006;48:477-482.
[117] Lebiedowska MK, Syczewska M. Invariant sway properties in children. *Gait Posture.* 2000;12:200-4.
[118] Williams HG, Fisher JM, Tritschler KA. Descriptive analysis of static postural control in 4, 6, and 8 year old normal and motorically awkward children. *Am J Phys Med.* 1983; 62:12–26.
[119] Heitkamp HC, Horstmann T, Mayer F, Weller J, Dickhuth HH. Gain in strength and muscular balance after balance training. *Int J Sports Medicine.*2001; 22:285-90.
[120] Liao CH, Yang YH, Chiang BL. Systemic lupus erythematosus with presentation as vertigo and vertical nystagmus: report of one case. *Acta Paediatr Taiwanica.* 2003;44:158-60.
[121] Bosco C, Komi PV. Influence of aging on themechanical behavior of leg extensor muscles. *Eur J Appl Physiol.* 1980;45:209-19.
[122] Hytonen M, Pyykko I, Aalto H, Starck J. Postural control and age. *Acta Otolaryngol.* 1993;113:119-22.
[123] Granacher U, Gollhofer A. Is there an association between variabes of postural control and strength in prepubertal children? *J Strength Cond Res.* 2012;26:210-6.
[124] Massion J. Movement, posture and equilibrium: interaction and coordination. *Prog Neurobiol.* 1992;38:35-56.
[125] Prioli AC, Cardozo AS, de Freitas Júnior PB, Barela JA. Task demand effects on postural control in older adults. *Hum Mov Sci.* 2006;25:435-46.
[126] Sahli S, Ghroubi S, Rebai H, Chaâbane M, Yahia A, Pérennou D, Elleuch MH. The effects of circus activity training on postural control of 5-6-year-old chikdren. *Science & Sports.* 2013;28:11-16.
[127] Adams JA. Historical review and appraisal of research on the learning, retention, and transfer of human motor skills. *Psychol Bull.*1987:101:41-74.
[128] Vuillerme N, Nougier V. Attentional demand for regulating postural sway: the effect of expertise in gymnastics. *Brain Res Bull.* 2004;63:161-65.
[129] Kioumourtzoglou E, Derri V, Mertzanidou O, Tzetzis G. Experience with perceptual and motor skills in rhythmic gymnastics. *Percept Mot Skills.*1997;84:1363-72.
[130] Robertson S, Collins J, Elliot D, Starkes J. The influence of skill and intermittent vision on dynamic balance. *J Mot Behav.* 1994;26:333-9.
[131] Henry FM. Specifcity versus generality in learning motor skill. In: Brown RC, Kenyon GS (eds). Classical studies on physical activity. Englewood Cliffs (NJ): Prentice Hall; 1968. p. 330-41.
[132] Simmons RW. Sensory organization determinants of postural stability in trained ballet dancers. *Int J Neurol.* 2005;115:87-97.
[133] Winter DA, Patla AE, Prince F, Ischac M, Gielo-Perczak F. Stiffness control of balance in quiet standing. *J Neurophysiol.* 1998;80:1211-21.

[134] Golomer E, Dupui P, Monod H. The effects of maturation on self-induced dynamic body sway frequencies of girls performing acrobatics or classical dance. *Eur J Appl Physiol Occup Physiol.* 1997;76:140-4.
[135] Crémieux J, Mesure S, Amblard B. Does the role of vision increase with the difficulty of the postural task? In: Taguchi K, Igarashi M, Mori S, editors. Vestibular and neural front. Pro-ceedings of the 12th international symposium on posture and gait, international congress series V, 1070. Amsterdam: Elsevier Science; 1994. p. 267-70.
[136] Olivier I, Palluel E, Nougier V. Effects of attentional focus on postural sway in children and adults. *Exp Brain Res.* 2008;185:341-5.

In: Posture: Types, Exercises and Health Effects
Editor: Sarah A. Curran
ISBN: 978-1-63117-252-6

Chapter 9

The Role of Unstable Shoe Constructions for the Improvement of Postural Control

Andreia S. P. Sousa*[1], *PhD
and João Manuel R. S. Tavares*,[2,*] *PhD
[1]Escola Superior da Tecnologia de Saúde do Porto, Área Científica de Fisioterapia, Centro de Estudos de Movimento e Actividade Humana, Vila Nova de Gaia, Portugal
[2]Instituto de Engenharia Mecânica e Gestão Industrial, Departamento de Engenharia Mecânica, Faculdade de Engenharia, Universidade do Porto, Porto, Portugal

Abstract

Postural control has been defined as the control of the body's position in space for the purposes of balance and orientation. Given the mechanical instability of the human body system, the neural process involved in stability organization and body orientation in space is necessary almost all motor actions. To manage movement variability, the postural control system presents a high adaptability in response to changing task and environment demands. The main sensory systems contribute to the development of an internal representation of body posture that is continuously updated based on multisensory feedback and is used to forward commands to control body position in space. For example, understanding the importance of proprioceptive information, and how ankle muscles can influence changes of support and stability that could improve postural control is of significant relevance. The purpose of this chapter is to review and discuss the short and long term influence of wearing an unstable shoe construction on postural control. The review provides instinctive knowledge that can be used during rehabilitation to improve motor performance. It also aims to provide a significant insight

* Corresponding author: tavares@fe.up.pt

into areas that have been dedicated to the implementation of preventive measures, such as ergonomy.

Keywords: Postural control, unstable shoe construction, reorganization, ankle muscles

Introduction

Postural control is an essential function in daily activities, which varies as a result of the task-individual-environment interaction. [1, 2] Indeed, considering that the body centre of mass (CoM) is located at two-thirds of body height above the ground, the human body is an inherently unstable system that provides a particularly challenging balance task to the central nervous system (CNS). [3] In turn, the CNS has to manage the redundancy of degrees of freedom resulting from the large number of muscles and joints involved to create flexible synergies according to task specificities. [4-9] The neural process involved in postural control is necessary for all dynamic motor actions. [10] In fact, most voluntary movement induces a postural perturbation because of dynamic, inter-segmental forces, and also shifts of the CoM. Therefore, voluntary movements may be considered self-inflicted postural perturbations that may be predicted, to a certain degree, by the CNS, which adjusts the activity of postural muscles both prior to the actual perturbation and in response to it. [11]

Studies on balance and posture during quiet or perturbed standing have identified the dominance of ankle muscles in the anteroposterior direction (Figure 1) and hip abductor/adductor muscles in the mediolateral direction. [3, 12-15] During quiet standing, postural sway results from a combination of inherent fluctuations in the musculoskeletal system, cardiac and respiratory variations, and neural activity. However, it has also been suggested that postural sway serves as an exploratory behavior for the stimulation of somatosensory and vestibular pathways to provide sensory information for increased postural control. [16, 17] By reconciling these two points of view, it is likely that sway characteristics result from an interaction between physiological states, the environment, and the implicit and explicit goals of the current task. [1, 18]

Under normal conditions during standing on a rigid floor, the postural control system elaborates the reference position using information about the relative positions of body segments, muscular torques and interaction with the base of support, taking into account the energy cost of standing and demands for stability and orientation. [19-21] Body alignment can minimise the effect of gravitational forces that tend to pull-off the CoM from the base of support, and muscle tone, i.e., intrinsic stiffness of the muscles, background muscle tone and postural tone, keeps the body from collapsing due to gravity. [1] When postural conditions change, the CNS must identify and selectively focus the most reliable sensory inputs to provide optimal control. As a result of this weighting of afferent input, muscle forces can be produced to control the CoM efficiently to maintain a good equilibrium. [22] Much of the research has focused in quantifying the human response to balance system perturbation in a different number of ways, such as displacement of support surface, predictable and unpredictable external perturbations and internal perturbations. Several factors have been shown to influence postural control responses, but only at an immediate level. Taking into account the high adaptability of the CNS [23] in response to changing task and environment demands, [1] it is important to understand the long-term influence of these changes of afferent information that could be beneficial to postural control. Given the importance of the ankle joint in upright postural control, [24] the influence of an unstable support base in postural control is of significant relevance.

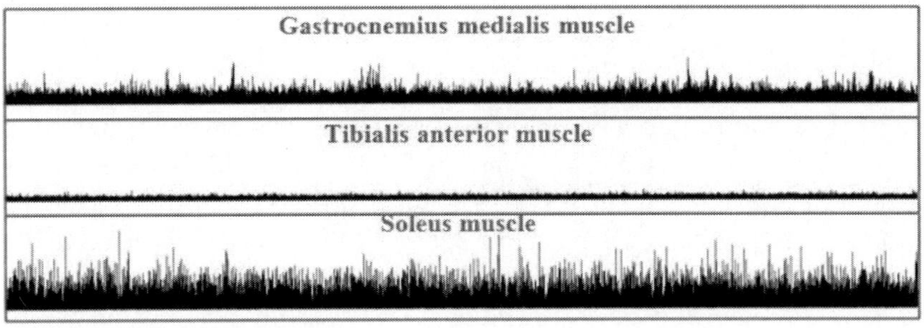

Figure 1. Muscle activity of ankle plantarflexors (soleus and gastrocnemius medialis muscles) and tibial anterior muscle during 60s of upright standing.

Neuromuscular Changes When Standing on an Unstable Support Base

When standing on an unstable support base the new postural requirements lead to postural control reorganisation through increased central drive [25, 26] associated with augmented gamamotoneuron activity leading to higher sensitivity of the muscle spindles [27] and higher muscle co-contraction. [28] However, anticipatory postural adjustments (APA) have been shown to decrease not only in very stable conditions, [29] but also in very unstable conditions. [30] Since the need for stabilising posture is diminished in stable conditions, the requirement for APA is also reduced. [29] Also, APA could themselves be a potential source of perturbation in case of unstable posture, and as such they are also reduced not to additionally destabilise posture. [30] Hence, it seems that the permanence in an unstable support base could, up to a certain level of instability, improve postural control.

Unstable Shoe Construction and Postural Control

Short-Term Postural Control Reorganisation in Response to an Unstable Shoe Construction

During quiet stance on a flat, stable platform individuals sway slightly, and the body oscillates around the ankle-joint axis like an inverted pendulum. [31] When standing on a see-saw (Figure 2), humans project the CoM onto the see-saw´s point of contact with the floor. [32]

There is an increased postural sway when wearing rocking shoes, [33] Table 1, expressed by changes in most of the representative centre of pressure (CoP) displacement parameters, [34] reflecting a situation of high postural demand, which associated to a higher active control at the ankle level [33, 35] would lead to an increased proprioceptive acuity provided by ankle muscles [36] as a result of a higher fusimotor drive. [27, 37, 38] This aspect is supported by recent studies that demonstrated a decrease of onset timing of ankle plantarflexors in response to an external forward perturbation under the condition of higher support instability. [39, 40] The relation between an increase of proprioceptive acuity and the higher fusimotor drive of ankle plantarflexors (gastrocnemius and soleus) during standing in an unstable support base is acceptable as length signals coming from the less adaptable spindles secondaries provide the CNS with an appropriate input for detecting low-frequency displacements occurring mainly about the ankle [21] and for assisting ankle muscle reflex responses. [41] This interpretation is valid if we consider the studies arguing that ankle plantarflexors are the main responsible for proprioceptive information signaling changes in body position. [42] Recently, based on the fact that during quiet standing reciprocal inhibition could be more important than autogenic stretch reflex, importance has been given to the role of un-modulated muscles crossing the joint in parallel with the active agonist. [43] Under this perspective, the use of unstable shoe construction (Figure 2), not affecting tibialis anterior muscle activity, [35, 39] would be also associated with increased acuity of proprioceptive information provided by the un-modulated muscles, since no differences were observed in antagonist muscles activity in response to the unstable support base condition. In summary, it can be suggested that the higher performance of the postural system when using unstable shoes is associated with an increased proprioceptive acuity related to higher motor drive of ankle plantarflexors spindles and/or to a consequent lower motor drive over muscle spindles of the antagonist.

Table 1. Representation of the variation of CoP displacement parameters obtained during standing with an unstable shoe construction in relation to a stable condition

Unstable shoe condition vs stable condition			
CoP displacement	**CoP root mean square**	**CoP mean velocity**	**CoP area**
↑	↑	↑	↑

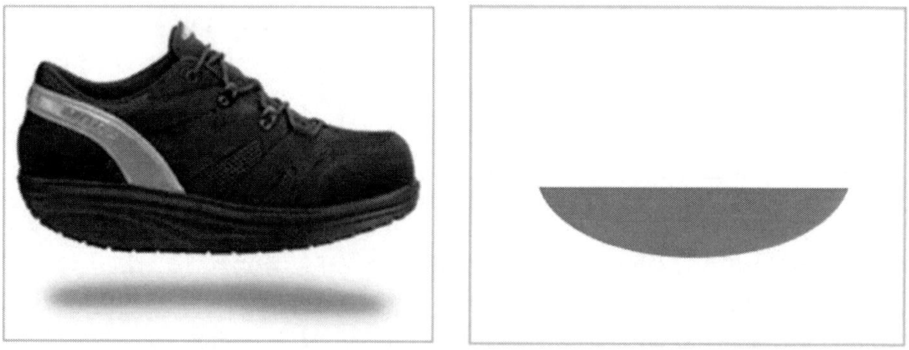

Figure 2. Illustration of a kind of unstable shoe construction and a schematic representation of the see-saw adopted in various studies to induce an unstable support base.

Studies about postural control when standing on a see-saw have devaluated the role of the information provided by muscle spindles, arguing that the higher changes in orientation provided by this kind of support surfaces makes it difficult to use proprioceptive information about the relative positions of successive links of the kinematic chain for the reconstruction of a body's position internal representation. [32, 44] As a consequence, the authors gave more importance to the information provided by the Golgi organ tendon. [32] However, it should be noted that the format of the unstable shoes commercialized is different from the see-saw used in previous studies, leading to lower levels of changes in segment orientation at the ankle joint, justifying the maintenance of the role of muscle spindles in this condition. The influence of the information provided by cutaneous afferents of the feet should be also questioned, considering its importance in standing postural control. [45, 46] However, to date, there is no evidence supporting this possibility.

In association with the previous paragraph, standing in an unstable support base is associated with increased reciprocal activation at the leg, thigh segments and muscle group levels, and lower co-activation levels. [39] Reciprocal activation of the agonist-antagonist set is present when the subject is uncertain of the task to be performed, when a quick compensatory force contraction is perceived to be required, [47] and in joints involving joint movement. [48] Studies using see-saws and continuous rotating platforms have indicated higher joint movement associated with the need of making see-saw rotation to place the support under the CoM. [32, 44] However, because the degree of perturbation induced by commercial unstable shoes is lower than that applied in the studies mentioned, it would be expected higher co-activation levels in the unstable support base condition, as co-activation has been described as the most robust strategy to counteract perturbations [49-52] by increasing joint stiffness. [53-57] However, the results obtained in recent studies [33, 39] indicate that wearing unstable shoes leads the postural control system to rely more on reflex feedback more than on stiffness increase to compensate for the decrease of stability, which has been demonstrated to be more efficient and accurate, but also more challenging for the postural control system. [58-64] This synergy pattern selection in association with the changes occurred in the magnitude of ankle muscles activity only [35] indicates that this kind of unstable shoe leads to instability levels that are perfectly managed by the CNS as a low perturbation. [65, 66] In fact, the postural performance is not modified when exposed to an external perturbation. [39, 40]

The higher demand over postural control in an unstable support base led to changes in the neuromusculoskeletal system, mainly at the ankle joint, leading also to positive effects over the venous return. [35] This could be related to the increased activity of the gastrocnemius muscle, to the kind of contraction imposed, as well as to the increased reciprocal activation.

Long-Term Postural Control Reorganisation in Response to an Unstable Shoe Construction

After prolonged standing in an unstable support base, the individual muscle activity, as well as the relation between agonist and antagonist, are close to the necessary for a stable support base condition. [33, 35, 40] However, exceptions were observed at the thigh level as reciprocal activation remained higher in the unstable condition [33, 40] associated to an increased biceps femoris activity in automatic and voluntary compensatory response and a decrease of gastrocnemius medialis activity. [40] These findings suggest a transfer of postural control synergy for the thigh which has been reported as more beneficial to optimise postural stability. [13, 19, 67-69] In fact, CoP parameters indicate higher performance by the postural control system in compensatory responses after prolonged wearing of unstable shoes, [40] and the relation between CoP and CoM [33] indicates a higher performance and efficiency [70, 71] of the supraspinal process, as well as of the action of spinal reflexes and/or of the intrinsic mechanical properties of muscles and joints. [72] Bearing in mind the important role of the information provided by group II fibers in postural control during standing, [41, 73-77] it can be hypothesised, based on the findings presented by Sousa et al., [40] that decreased gastrocnemius activity after prolonged wearing of unstable shoes could also be related to a possible decrease of medium latency responses. This reduction could be because of the relationship to group II afferences, being more influenced by the postural set and being the only that have a stabilising effect during perturbation of stance. [78, 79] The maintenance of decreased values of ankle muscle latency, even after prolonged unstable shoe wearing, [40] can be related to a remaining instability effect expressed by the higher activity of total agonist activity and higher values of CoP displacement in the unstable support base condition. [33]

From the findings and observations presented in this chapter, [33, 40] it is evident that prolonged wearing of unstable shoes leads to improved postural performance and efficiency. Also, even after adaptation by the postural control system, the venous return is higher in the unstable support base condition than in the control condition. [35] The results obtained by Sousa et al. [40] suggest that the main factor responsible for increased venous return is related to the kind of muscle contraction of plantarflexors, since dynamic muscle contractions favour venous circulation. This conjecture is sustained by higher CoP displacement values observed after prolonged wearing of unstable shoe [33] and by studies on see-saws. [32, 44]

The impact of postural control strategies on venous return is of significant relevance in healthy subjects that are in risk of developing venous insufficiency, but also in subjects with venous insufficiency.

Chronic venous insufficiency explains those manifestations of venous disease resulting from ambulatory venous hypertension, which is associated with failure of the lower extremity muscle pumps due to outflow obstruction, musculo-fascial weakness, loss of joint motion or valvular failure. [80-83] Although the effect of an unstable support base has been explored only in healthy subjects, beneficial results would be expected in subjects with venous insufficiency, as balance training has been demonstrated to promote improved postural control performance both in healthy subjects [84-90] and in subjects with ankle muscles impairment. [87, 91] This is of significant relevance since it has been demonstrated that calf muscle strengthening exercises restore the pumping ability of the calf muscle and improve the haemodynamic performance in limbs with active ulceration subsequent to severe venous valvular and calf muscle pump impairment. [92-94]

Conclusion

This review chapter has discussed the role of instability imposed by rocking (unstable) shoes and how it may lead to positive effects over the postural control system. This effect appears to be related to higher acuity muscle spindles at the immediate level and after a long-term exposure. The changes occurred in terms of ankle muscle activity in response to an unstable support base favour venous circulation even after training adaptation by the neuromuscular system. Considering the review presented in this chapter, the use of unstable shoe constructions should be encouraged in rehabilitation as a strategy to improve postural control performance.

Acknowledgments

The first author would like to thank the PhD grant and the support and contribution from Instituto Politécnico do Porto (IPP) and Escola Superior de Tecnologia da Saúde (ESTSP), in Portugal.

References

[1] Shumway-Cook A, Woolacott M. *Motor control: translating research into clinical practice*. 3rd ed. Philadelphia: Lippincott Williams & Wilkins; 2007.

[2] Allum JHJ, Bloem BR, Carpenter MG, Hulliger M, Hadders-Algra M. Proprioceptive control of posture: a review of new concepts. *Gait Posture*. 1998;8:214-42.

[3] Winter DA. Human balance and posture control during standing and walking. *Gait Posture*. 1995;3:193-214.

[4] Allum JHJ, Honegger F, Schicks H. Vestibular and proprioceptive modulation of postural synergies in normal subjects. *J Vestibular Research*. 1993;3:59-85.

[5] Allum JHJ, Honegger F, Schicks H. The influence of a bilateral peripheral vestibular deficit on postural synergies. *J Vestibular Research*. 1994;4:49-70.

[6] Diener HH, FB; Nashner, LM. Influence of stimulus parameters on human postural responses. *J Neurophysiology*. 1988;59:1888-905.

[7] Horak B, Nashner L. Central programming of postural movements: adaptation to altered support-surface configurations. *J Neurophysiology*. 1986;55:1369-81.

[8] Keshner EW, MH; Debu, B. Neck, trunk and limb muscle responses during postural perturbations in humans. *Experimental Brain Research*. 1988;71:455-66.

[9] Macpherson J. Changes in postural strategy with inter-paw distance. *J Neurophysiology*. 1994;71:931-40.

[10] Massion J. Postural control systems in developmental perspective. *Neuroscience Biobehavioral Review*. 1998;22:465-72.

[11] Arruin A. The organization of anticipatory postural adjustments. *J Automatic Control*. 2002;12:31-7.

[12] Fitzpatrick R, Taylor J, McCloskey D. Ankle stiffness of standing humans in response to imperceptible perturbation: reflex and task-dependent components. *J Physiology*. 1992;454:533-47.

[13] Kuo AZ, FE. Human standing posture: multi-joint movement strategies based on biomechanical constraints. *Progress in Brain Research*. 1993;97:349-58.

[14] Johanson R. System modeling and identification. Englewood Cliffs: *Prentice Hall;* 1993.

[15] Winter DA, Prince F, Frank J, Powell C, Zabjek K. Unified theory regarding A/P and M/L balance in quiet stance. *J Neurophysiology*. 1996;75:2334-43.

[16] Riley MA, Balasubramaniam R, Turvey MT. Recurrence quantification analysis of postural fluctuations. *Gait & Amp; Posture*. 1999;9:65-78.

[17] Riley MA, Wong S, Mitra S, Turvey MT. Common effects of touch and vision on postural parameters. *Experimental Brain Research*. 1997;117:165-70.

[18] Rothwell J. Control of human voluntary movement. 2nd ed. London: *Chapman & Hall;* 1994.

[19] Horak F, Nashner L, Diener H. Postural strategies associated with somatosensory and vestibular loss. *Experimental Brain Research.* 1990;82:167-77.

[20] Dietz V. Human neuronal control of automatic functional movements: interaction between central programs and afferent input. *Physiological Reviews.* 1992;72:33-69.

[21] Gurfinkel VS, Ivanenko YP, Levik YS, Babakova IA. Kinesthetic reference for human orthograde posture. *Neuroscience.* 1995;68:229-43.

[22] Carver S, Kiemel T, Jeka J. Modeling the dynamics of sensory reweighting. *Biological Cybernetics.* 2006;95:123-34.

[23] Winter DA. Kinematic and kinetic patterns in human gait: Variability and compensating effects. *Human Movement Sci.* 1984;3:51-76.

[24] Fitzpatrick R, Douglas K, McCloskey D. Stable human standing with lower-limb muscle afferents providing the only sensory input. *J Physiology.* 1994;2536:395-403.

[25] Ivanenko YP, Talis VL, Kazennikov OV. Support stability influences postural responses to muscle vibration in humans. *European J Neuroscience.* 1999;11:647-54.

[26] Gavrilenko T, Gatev N, Dimitrova D. Somatosensory evoked potentials during standing posture on different support conditions. *Homeostasis.* 1995;33:39-46.

[27] Ribot-Ciscar E, Hospod V, Roll J-P, Aimonetti J-M. Fusimotor drive may adjust muscle spindle feedback to task requirements in humans. *J Neurophysiology.* 2009;101:633-40.

[28] Dietz V, Mauritz K, Dichgans J. Body oscillations in balancing due to segmental stretch reflex activity. *Experimental Brain Research.* 1980;40:89-95.

[29] Nardone A, Schieppati M. Postural adjustments associated with voluntary contraction of the leg muscles in standing man. *Experimental Brain Research.* 1988;69:469-80.

[30] Arruin A, Forrest W, Latash M. Anticipatory postural adjustments in conditions of postural instability. *Electroencephalography Clinical Neurophysiology.* 1998;109: 350-9.

[31] Winter DA, Patla A, Prince F, Ishac M, Gielo-Perczak K. Stiffness control of balance in quiet standing. *J Neurophysiology.* 1998;80:1211-21.

[32] Ivanenko Y, Levik Y, Talis V, Gurfinkel V. Human equilibrium on unstable support: the importance of feet-support interaction. *Neuroscience Letters.* 1997;235:109-12.

[33] Sousa ASP, Macedo R, Santos R, Silva A, Tavares JMR. The influence of wearing unstable shoes on upright standing postural control. *Medical Engineering & Physics.* 2013;(submitted).

[34] Rocchi L, Chiari L, Cappello A. Feature selection of stabilometric parameters based on principal component analysis. *Med Biol Eng Comput.* 2004;42:71-9.

[35] Sousa ASP, Tavares JMRS, Macedo R, Rodrigues AM, Santos R. Influence of wearing an unstable shoe on thigh and leg muscle activity and venous response in upright standing. *Applied Ergonomics.* 2012;43:933-9.

[36] Gandevia SC, McCloskey DI, Burke D. Kinaesthetic signals and muscle contraction. *Trends in Neurosciences.* 1992;15:62-5.

[37] Gorassini M, Prochazka A, Taylor JL. Cerebellar ataxia and muscle spindle sensitivity. *J Neurophysiology.* 1993;70:1853-62.

[38] Gurfinkel VS, Popov KK, B.N. S. The support input as a reference for postural control. In: Woolacott M, Horak F (eds). Posture and gait: Control mechanisms. Oregon: University of Oregon Books; 1992. p. 182-6.

[39] Sousa ASP, Macedo R, Santos R, Tavares JMR. The influence of wearing unstable shoes on compensatory control of posture. *Human Movement Sci.* 2013; in press.

[40] Sousa ASP, Macedo R, Silva A, Santos R, Tavares JMR. Influence of long-term wearing of unstable shoes on compensatory control of posture: An electromyography-based analysis. *Gait Posture.* 2014;39:98-104.

[41] Schieppati M, Nardone A, Siliotto R, Grasso M. Early and late stretch responses of human foot muscles induced by perturbation of stance. *Experimental Brain Research.* 1995;105:411-22.

[42] Loram I, Maganaris C, Lakie M. Active, non-spring-like muscle movements in human postural sway: how might paradoxical changes in muscle length be produced? *J Physiology.* 2005;564:283-93.

[43] Di Giulio I, Maganaris C, Baltzopoulos V, Loram I. The proprioceptive and agonist roles of gastrocnemius, soleus and tibialis anterior muscles in maintaining human upright posture. *J Physiology.* 2009;587:2399-416.

[44] Ivanenko Y, Talis V. Effect of surface support stability on the postural vibration reactions in man. *Human Physiology.* 1995;21:116-24.

[45] Roll R, Kavounoudias A, Roll J-P. Cutaneous afferents from human plantar sole contribute to body posture awareness. *NeuroReport.* 2002;13:1957-61.

[46] Wright W, Ivanenko Y, Gurfinkel V. Foot anatomy specialization for postural sensation and control. *J Neurophysiology.* 2012;107:1513-21.

[47] De Luca C, Mambrito B. Voluntary control of motor units in human antagonist muscles: coactivation and reciprocal activation. *J Neurophysiology.* 1987;58:525-42.

[48] Lavoie B, Devanne H, Capaday C. Differential control of reciprocal inhibition during walking versus postural and voluntary motor tasks in humans. *J Neurophysiology.* 1997;78:429-38.

[49] Milner T. Adaptation to destabilizing dynamics by means of muscle cocontraction. *Experimental Brain Research.* 2002;143:406-16.

[50] Osu R, Gomi H. Multijoint muscle regulation mechanisms examined by measured human arm stiffness and EMG signals. *J Neurophysiology.* 1999;81:1458-68.

[51] Humphrey D, Reed D. Separate cortical systems for control of joint movement and joint stiffness: reciprocal activation and coactivation of antagonist muscles. *Advances Neurology.* 1983;39:347-72.

[52] Damm L, McIntyre J. Physiological basis of limb-impedance modulation during free and constrained movements. *J Neurophysiology.* 2008;100:2577-88.

[53] Feldman A. Superposition of motor programs: 1. Rhythmic forearm flexion in man. *Neuroscience.* 1980;5:81-90.

[54] Joyce GC, Rack PMH, Westbury DR. The mechanical properties of cat soleus muscle during controlled lengthening and shortening movements. *J Physiology.* 1969;204:461-74.

[55] Nichols TR, Houk JC. Improvement in linearity and regulation of stiffness that results from actions of stretch reflex. *J Neurophysiology.* 1976;39:119-42.

[56] Serres SJ, Milner TE. Wrist muscle activation patterns and stiffness associated with stable and unstable mechanical loads. *Experimental Brain Research.* 1991;86:451-8.

[57] Milner T, Cloutier C, Leger A, Franklin D. Inability to activate muscles maximally during cocontraction and the effect on joint stiffness. *Experimental Brain Research.* 1995;107:293-305.

[58] Friedli WG, Hallett M, Simon S. Postural adjustments associated with rapid voluntary arm movements 1. Electromyographic data. *J Neurology, Neurosurgery & Psychiatry.* 1984;47:611-22.

[59] Hong D, Corcos D, Gottlieb G. Task dependent patterns of muscle activation at the shoulder and elbow for unconstrained arm movements. *J Neurophysiology.* 1994;71:1261-5.

[60] Latash M, Aruin A, Neyman I, Nicholas J. Anticipatory postural adjustments during self-inflicted and predictable perturbations in Parkinson's disease. *J Neurology, Neurosurgery & Psychiatry.* 1995;58:326-34.

[61] Hogan N. Adaptive control of mechanical impedance by coactivation of antagonist muscles. *IEEE Trans Autom Control.* 1984;29:681-90.

[62] Aruin A, Almeida G. A coactivation strategy in anticipatory postural adjustment in persons with Down syndrome. *Motor Control.* 1997;2:178-97.

[63] Garland SJ, Stevenson TJ, Ivanova T. Postural responses to unilateral arm perturbation in young, eldery, and hemiplegic subjects. *Arch Physical Med Rehab.* 1997;78:1072-7.

[64] Massion J, Ioffe M, Schmitz C, Viallet F, Gantcheva R. Acquisition of anticipatory postural adjustments in bimanual load-lifting task: normal and pathological aspects. *Experimental Brain Research.* 1999;128:229-35.

[65] Santos M, Aruin A. Effects of lateral perturbations and changing stance conditions on anticipatory postural adjustment. *J Electromyography Kinesiology.* 2009;19:532-41.

[66] Adkin A, Frank J, Carpenter M, Peysar G. Postural control is scaled to level of postural threat. *Gait Posture. 2000*;12:87-93.

[67] Day BL, Steiger MJ, Thompson PD, Marsden CD. Effect of vision and stance width on human body motion when standing: implications for afferent control of lateral sway. *J Physiology.* 1993;469:479-99.

[68] Runge C, Shupert C, Horak F, Zajac F. Ankle and hip postural strategies defined by joint torques. *Gait Posture.* 1999;10:161-70.

[69] Yang JF, Winter DA, Wells RP. Postural dynamics in standing human. *Biological Cybernetics.* 1990;62:309-20.

[70] Prieto T, Myklebust J, Hoffmann R, Lovett E, Myklebust B. Measures of postural steadiness: differences between healthy young and elderly adults. Biomedical Engineering, *IEEE Transactions on.* 1996;43:956-66.

[71] Bennell K, Goldie P. The differential effects of external ankle support on postural control. *JOSPT.* 1994;20:287-95.

[72] Zatsiorsky VM, Duarte M. Instant equilibrium point and its migration in standing tasks: rambling and trembling components of the stabilogram. *Motor Control.* 1999;3:28.

[73] Corna S, Galante M, Grasso M, Nardone A, Schieppati M. Unilateral displacement of lower limb evokes bilateral EMG responses in leg and foot muscles in standing humans. *Experimental Brain Research.* 1996;109:83-91.

[74] Marchand-Pauvert V, Nicolas G, Marque P, Iglesias C, Pierrot-Deseilligny E. Increase in group II excitation from ankle muscles to thigh motoneurones during human standing. *Journal Physiology.* 2005;566:257-71.

[75] Morasso PG, Schieppati M. Can muscle stiffness alone stabilize upright standing? *Journal of Neurophysiology.1999*;82:1622-6.
[76] Nardone A, Grasso M, Giordano A, Schieppati M. Different effect of height on latency of leg and foot short- and medium-latency EMG responses to perturbation of stance in humans. *Neuroscience Letters.* 1996;206:89-92.
[77] Schieppati M, Nardone A. Medium-latency stretch reflexes of foot and leg muscles analysed by cooling the lower limb in standing humans. *J Physiology.* 1997;503:691-8.
[78] Nardone A, Giordano A, Corrá T, Schieppati M. Responses of leg muscles in humans displaced while standing. Effects of types of perturbation and of postural set. *Brain.* 1990;113:65-84.
[79] Jacobs JV, Horak FB. Cortical control of postural responses. *J Neural Transm.* 2007 2007/10/01;114:1339-48.
[80] Araki C, Back T, Padberg F, Thompson P, Jamil Z, Lee B. The significance of calf muscle pump function in venous ulceration. *J Vasc Surg.* 1994;20:872-7.
[81] Plate G, Brudin L, Eklof B, Jensen R, Ohlin P. Congenital vein valve aplasia. *World J Surg.* 1986;10:929-34.
[82] Nicolaides A, Hussein M, Szendro G, Christopoulos D, Vasdekis S, Clarke H. The relationship of venous ulceration with ambulatory venous pressure measurements. *J Vasc Surg.* 1993;17:414-9.
[83] Stewart J, Medow M, Montgomery L, McLeod K. Decreased skeletal muscle pump activity in patients with postural tachycardia syndrome and low peripheral blood flow. *Am J Heart Circulatory Physiology.* 2004;286:1216-22.
[84] Balogun J, Adesinasi C, Marzouk D. The effects of a woble board exercise training program on static balance performance and strength of lower extremity muscles. *Physiotherapy Canada.* 1992;44:23-30.
[85] Heitkamp H, Horstmann T, Mayer F, Weller J, Dickhurth H. Gain in strength and muscular balance after balance training. *Int J Sports Med.* 2001;22:285-90.
[86] Hoffman M, Payne V. The effect of proprioceptive ankle disk training on healthy subjects. *JOSPT.* 1995;21:90-3.
[87] Rozzi S, Lephart S, Sterner R, Kuligowski L. Balance training for persons with functionally unstable ankles. *JOSPT.* 1999;29:478-86.
[88] Waddington G, Adams R. The effect of a 5-week wobble-board exercise intervention on ability to discriminate different degrees of ankle inversion, barefoot and wearing shoes: a study in healthy eldery. *J Am Geriatrics Society.* 2004;52:573-6.
[89] Waddington G, Seward H, Wrigley T, Lacey N, Adams R. Comparing wobble board and jump-landing training effects on knee and ankle movement discrimination. *J Sci Med Sport.* 2000;3:449-59.
[90] Wester J, Jesperson S, Nielsen K, Neumann L. Wobble board training after partial sprains of the lateral ligaments of the ankle: a prospective randomized study. *JOSPT.* 1996;23:332-6.
[91] Mattacola C, Lloyd J. Effects of a 6-week strength and proprioceptive training program on measures of dynamic balance: a single-case design. *J Athletic Training.* 1997;32:127-35.
[92] Yang D, Vandongen Y, Stacey M. Effect of exercise on calf muscle pump function in patients with chronic venous disease. *Brit J Surg.* 1999;86:338-41.

[93] Kan Y, Delis K. Hemodynamic effects of supervised calf muscle exercise in patients with venous leg ulceration. A prospective controlled study. *Arch Surg.* 2001;136:1364-9.

[94] Padberg F, Johnston M, Sisto S. Structured exercise improves calf muscle pump function in chronic venous insufficiency: a randomized trial. *J Vasc Surg.* 2004;39:79-87.

In: Posture: Types, Exercises and Health Effects
Editor: Sarah A. Curran
ISBN: 978-1-63117-252-6

Chapter 10

Efficacy of Modified Yoga Positions and Postural Chains Therapy for Spinal Pain Treatment

José Luís Pimentel do Rosário, [*] ***PT, PhD***
Federal University of São Paulo, São Paulo, Brazil

Abstract

Several types of musculoskeletal pathologies can be related to poor posture. This may occur because of joint stress and the consequent increase in muscle energy expenditure, thereby creating tensional stress in muscles, joints, ligaments, fascia, bones and even arteries and peripheral nerves. *Postural chains therapy* (PCT) is a technique that uses modified yoga positions to treat postural pathologies, among others, fitting these yoga positions into the Mézières concept of muscular chains. The body works as a whole chain and these chains can be subdivided into the following: posterior chain; inspiratory chain; hip adductor chain; arm internal rotator chain; arm adductor chain and arm elevator-abductor chain. This chapter presents a study exploring the efficiency of a single session of two modified yoga positions with 200 subjects and their 235 spinal pain-related complaints. The subjects were divided into two groups: The yoga group that received a treatment of two 15 minutes postures and the control group that received a placebo treatment of two 15 minutes relaxation in the supine position. All volunteers experienced some pain before treatment and were assessed before and after treatment using the analog pain scale. A score of 0 indicated no pain whereas 10 was the maximum degree of pain on the scale. The average analog pain scale values drop from 5.55 to 1.79 in the yoga group and from 5.42 to 4.64 in the control group. The difference before and after treatment was compared between the groups with a p-value = 0.0001, as measured by the Student's t-test. It is possible to conclude that one therapy session is effective in the treatment of various musculoskeletal problems.

[*] Contact email: joerospt@gmail.com

Keywords: Manual therapy, muscular chain, posture, yoga, analog scale, musculoskeletal, pain, treatment

Introduction

The word posture refers to the alignment and maintenance of body segments in certain positions such as standing or seated against gravity action. [1] In order to achieve this goal, some tonic muscles, which can maintain a certain level of contraction over a long period of time, use the bones as levers. When there is an unbalance between these muscles, poor posture takes over. If body segments are kept out of alignment for extended periods, some of the muscles involved are used in a shortened position as a consequence. [2, 3] These muscles are usually seen as strong, while their antagonists are taken to be elongated and weak, and this is one of the effects of poor posture.

Poor posture can compromise more than just muscles. In comparison, good posture creates the least amount of joint stress and requires the least amount of muscle activity. Consequently, it is the position of maximum efficiency. [4, 5] However, a deviation from optimal positioning is compensated by changes in the joint position which, in turn, must be maintained by an increase in muscle activity. [6, 7] Therefore, postural instability results in an excess of power consumption, muscle tension and joint stress. This change in joint position described by Lee, [4] in reference to the spine, is very similar to the concept of chiropractic subluxation. Generations of chiropractors have claimed that a large percentage of all diseases are caused by subluxation, involving not only the musculoskeletal system, but the neural system as well [9].

Focusing on muscle imbalances, these postural deviations can adversely affect muscular efficiency, predisposing individuals to pathologic musculoskeletal conditions. One of the major symptoms of postural change is pain. [9] In support of this observation, Moreira et al. [10] investigated the use of physical therapy to correct their shoulder posture and to reduce the pain caused by postural abnormalities in a group of adult females. The patients were radiographed one week after therapeutic discharge. The postural improvement was evidenced by the retraction of the shoulder and a reduction in the pain experienced.

In 1947, kinesiotherapist Françoise Mézières, the originator of therapies based in muscular chain [11, 12] stated that human muscles are completely inter-related and demonstrated that there is not just a single muscle that causes a poor posture. Instead, it is related to chains of muscles that can end up causing a pathology in a specific place from a generalized tension. Therefore, a localized muscular action provokes reactions at a distance, underlining that the root of the problem can be distant from where the patient feels pain. [13] More recently, Hoppenfeld (14) in accordance with Mézières theories, highlighted in his book on orthopedic evaluation that referred pain in a joint can have its cause in adjacent joints.

The postural chains therapy (PCT) uses this postural concept observed and elaborated by Mézières in order to understand the posture. For the evaluation, PCT uses two yoga positions and five for the treatment. These yoga positions were modified aiming to stretch at the same time all the muscles of the chain to be worked. A number of existing scientific studies have approached PCT or similar techniques based on Meziérès muscular chains and modified yoga postures with assistance of a therapist. These studies have produced results in treating various musculoskeletal conditions. [15-23]. Considering the fact that poor posture is associated with joint positioning changes, and that this malposition can cause pain, the origin of which is far from the location.

The aim of this chapter is to present a study which set out to determine the efficiency of a single application of modified yoga positions. These yoga positions were modified aiming to stretch at the same time all the muscles of the posterior and anterior chains based on the work of Mézières. [24]

Methods

In total, 200 volunteers between 20 and 39-years-old participated in the study. The inclusion criterion for the subjects was to be currently experiencing some spinal musculoskeletal pain. Each subject could complain of more than one spinal pain. For this reason, the total number of complaints was 235. The exclusion criteria included any psychiatric or neurological disorders. Volunteers signed a statement of informed consent and were assessed and treated with muscular chain therapy (MCT), as described by Rosário, [7] or placed in a placebo group. The present study received approval from the Human Research Ethics Committee of the State University of the Center-West (UNICENTRO), under protocol number 289/2011.

The subjects were divided into two groups of 100 participants each: The MCT group that received MCT and the control group that received a placebo treatment, which consisted in relaxing while laying down in supine. The name of this position in sanskrit is shavasana, which means *"corpse pose"*. Both groups were treated by a physical therapist with MCT certification and knowledge in yoga with two postures during 15 minutes each, with a 2 minutes interval between them.

MCT Group Assessment and Comparison of Muscle Chains

The assessment of the anterior and posterior chains as described by Rosário [24] are explained below, and were performed to find the most compromised muscular chain. The chain presenting the highest compensation was the first to be treated.

MCT Treatment

Treatment consisted of two postures of 15 minutes each. Before treatment, the subjects were taught how to separate breathing by region: apical; lower ribs and diaphragmatic breathing, aimed at helping the maintenance of posture. The selection of posture was based on the assessment described above. If the therapist found more postural alterations in the posterior chain, two postures of the posterior chain were performed. If more deviations were found in the anterior chain, two postures of the anterior chain were performed. If the two chains exhibited similar deviations, the treatment included one posture for each chain. Both of the supine postures were performed according to Rosário [24] and are explained below.

Postural Chains Therapy

Muscular chains are formed by gravitational muscles that work synergistically in the same chain [very well explained by the theory of anatomy trains]. [24,25] The maintenance of the standing position against the action of gravity shapes these muscles and their connective tissue. [24-26] The concept of muscular chains is based on the observation that the shortening of a muscle creates compensation in the adjacent and also distant muscles. In this study the fascial treatment was not added in order to emphasize the muscles. Therefore, PCT based on the muscular chains theory, is a method of global static stretching that uses postural positions to stretch several muscle groups simultaneously rather than simply treating an isolated muscle. [1-24] There are basically two main muscle chains: the anterior and the posterior, which have strong similarities with the anatomy trains' superficial frontal line and superficial back line, [25] respectively.

The posterior chain (Figure 1) includes the following muscles: gastrocnemius and soleus; flexor hallucis brevis; flexor hallucis longus; short flexor of fingers and flexor digitorum longus; adductor hallucis; abductor hallucis; hamstrings (semitendinosus; semimembranosus and biceps femoris); popliteal; gluteus maximus. It also includes the following paraspinals: iliocostalis lumborum; iliocostalis thoracis; iliocostalis cervicis; longissimus thoracis; longissimus cervicis; longissimus capitis; spinalis thoracis; spinalis cervicis; spinalis capitis; semispinalis dorsi; semispinalis cervicis; semispinalis capitis; multifidus; rotatores; interspinales, and the intertransversarii. [26, 27]

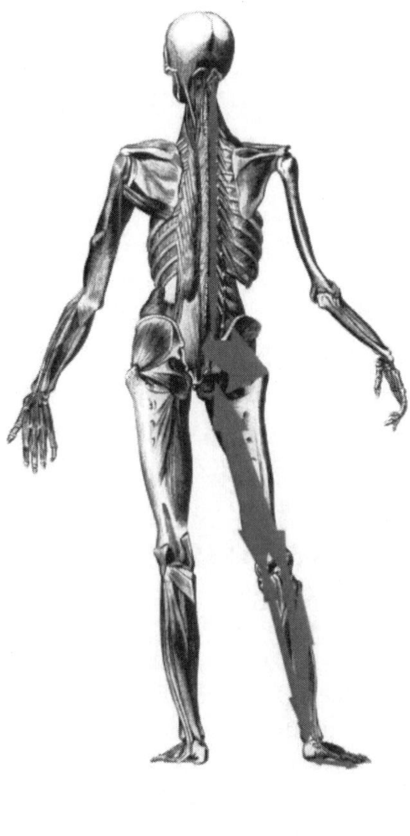

Figure 1. Posterior chain.

The anterior chain is divided in two: the inspiratory chain (Figure 2) and the hip internal rotator chain (Figure 3).

The *inspiratory chain* includes the following muscles: scalenus; sternocleidomastoid; pectoralis minor; intercostals and the diaphragm. [26, 27] The hip internal rotator chain includes the following muscles: iliacus; psoas minor; psoas major; adductor brevis; adductor longus; adductor magnus; gracilis; and pectineus. [26, 27]

The *arm chains* can be divided in three: the shoulder adductor chain that may be considered part of *anterior chain*; the *shoulder elevator-abductor chain* that may be considered part of the *posterior chain*; and the *arm internal rotator chain* which is neutral, because it is associated to both *anterior* and *posterior chains* and can also be stretched in *anterior* and *posterior* postures.

The *arm internal rotator chain* (Figure 4) includes the following muscles: pectoralis major; brachial biceps and brachialis; pronator teres and pronator quadratus; flexor digitorum profundus; flexor digitorum superficialis; flexor pollicis longus and adductor pollicis; abductor pollicis brevis; flexor pollicis brevis, and the opponens pollicis. [26, 27]

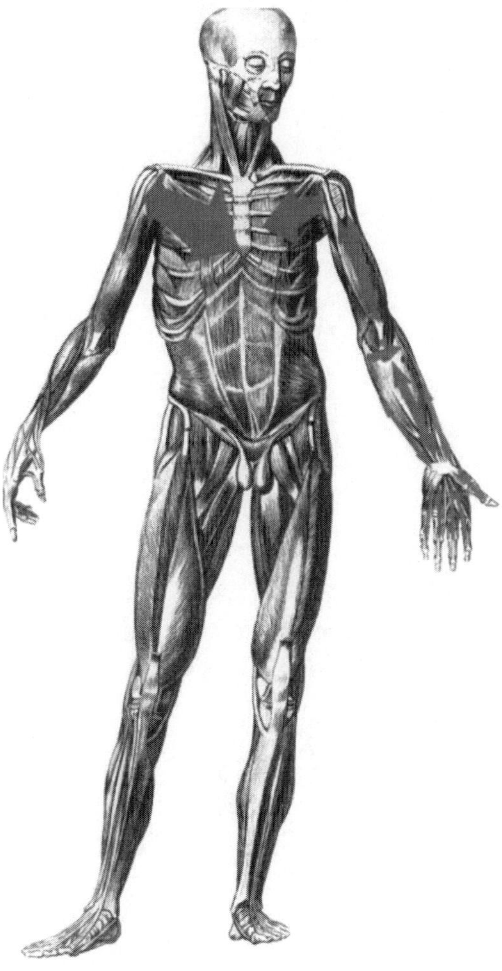

Figure 2. Inspiratory chain.

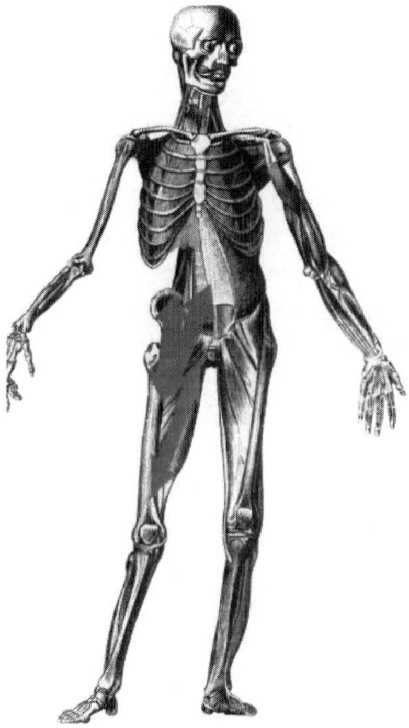

Figure 3. Hip internal rotator chain.

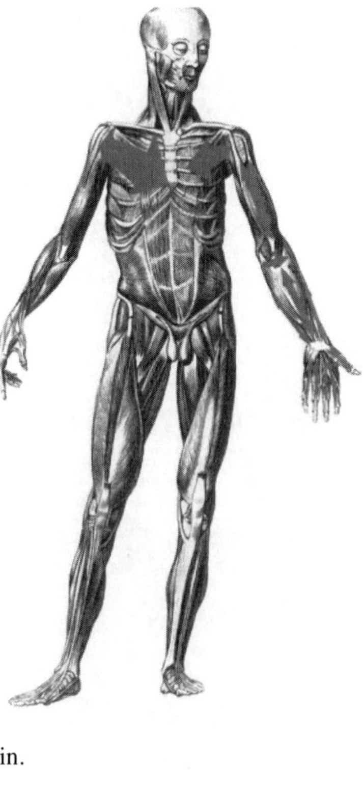

Figure 4. Arm internal rotator chain.

The *shoulder elevator-adductor chain* (Figure 5) includes the following muscles: subscapularis; deltoid, and the upper trapezius. The *shoulder adductor chain* (Figure 6) includes the following muscles: pectoral major; coracobraquialis, and the subscapularis. [26, 27]

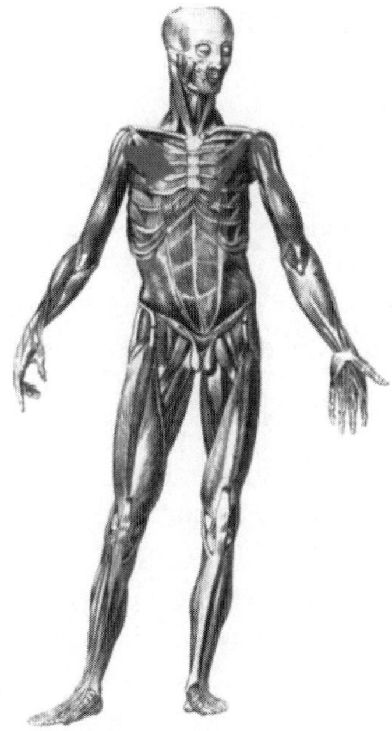

Figure 5. Shoulder elevator-adductor chain.

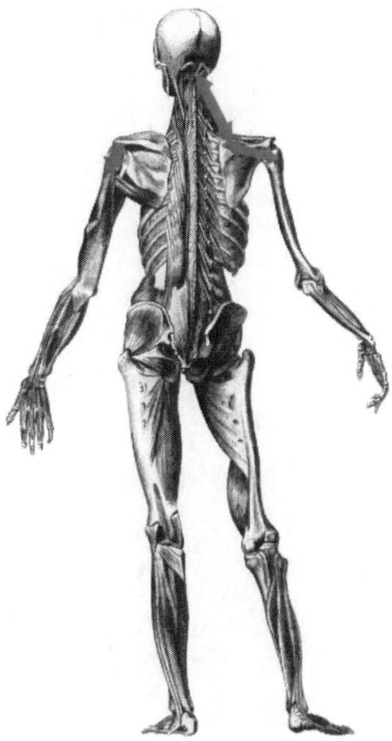

Figure 6. Shoulder adductor chain.

This muscular chain concept differs from the segmental stretch which treats each shortened muscle separately, usually those directly involved in the joint with decreased range of motion. PCT applies a long duration stretch, which lasts approximately 15 minutes per postural position while coupled with eccentric physical exertion for the posture maintenance, featuring an active form of stretching. [1-27] Clinically, the MCT has been efficient treating postural deviations and providing greater flexibility. [27]

Assessment with PCT is described according to Rosário et al. [24] (A – Assessment).

Examination of the Anterior Chain

The anterior muscles that can raise the lordosis are the diaphragm, the iliopsoas and the pubic adductors. They pull the lumbar spine forward, placing the pelvis in anteversion. Therefore, it was necessary to rectify the lumbar in order to test this chain. The tadasana – *mountain pose* – is used to assess the anterior chain (Figure 4). Patients who have either shortening or tension in this chain exhibit compensatory changes.

Assessment Steps

1) Place the individual standing with their heels together;
2) Rectify lumbar lordosis by retroversion;
3) Observe the compensation, which may be one of the following:
    - Leaning the torso back (Figure 7);
    - Bending the knees. A small flexion is normal (Figure 8);
    - Chest blocked in an inspiratory position;
    - Protrusion of the head and shoulders (Figure 9).

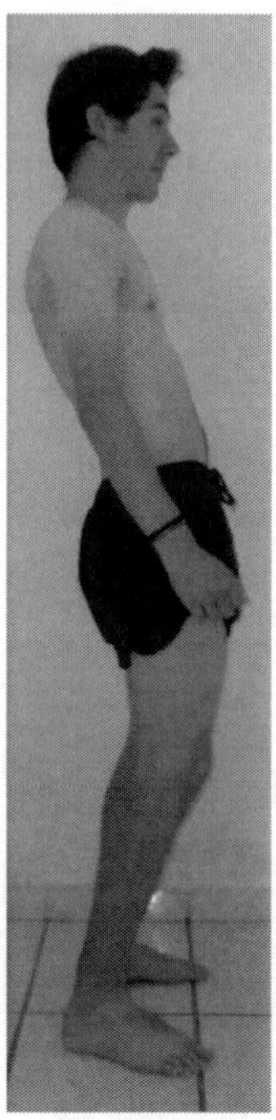

Figure 7. Compensation for the anterior chain test: leaning the torso back.

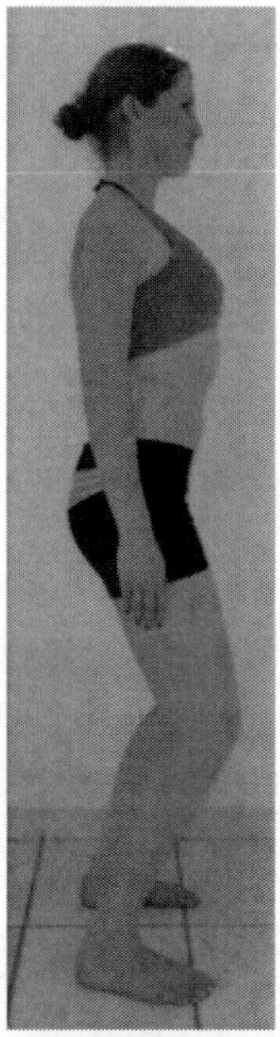

Figure 8. Compensation for the anterior chain test: bending the knees and inspiratory blocking.

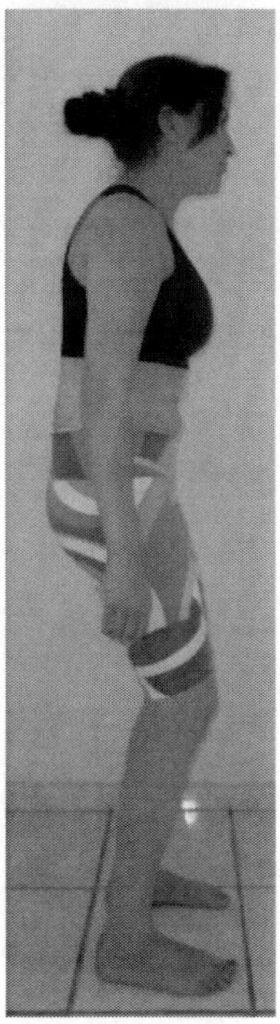

Figure 9. Compensation for the anterior chain test: head protrusion.

Examination of the Posterior Chain

The posterior spinal muscles flatten and push the lumbar spine back, leading to pelvis retroversion, with a tendency to keep the sacrum in a horizontal position. It is necessary to flex the trunk in order to assess this chain. The flexion alone can provide some clues about posture. Since this chain tends to invert the lumbar curve, causing a kyphosis, it is important to request a small lumbar lordosis with the hip flexion in order to put this chain in full tension. Consequently, the trunk will be raised but a loosening of hip flexion will be avoided. When it exists, shortening becomes obvious at this stage. The shaktyasana – *The shakty goddess pose* – is used to assess the posterior chain (Figure 10).

Assessment Steps

1) The subject must lean forward with the knees straight and the heels together. Note if the curvature of the spine and the spinous processes are all clearly visible.
2) To complete the test, align the lumbar spine, requesting a small lumbar lordosis.
3) A shortened individual will extend the trunk opening the hip angle of flexion (Figure 11). Another possible abnormality is the opening of the ankle angle (Figure 12), which is the angle formed between the foot and the tibia.

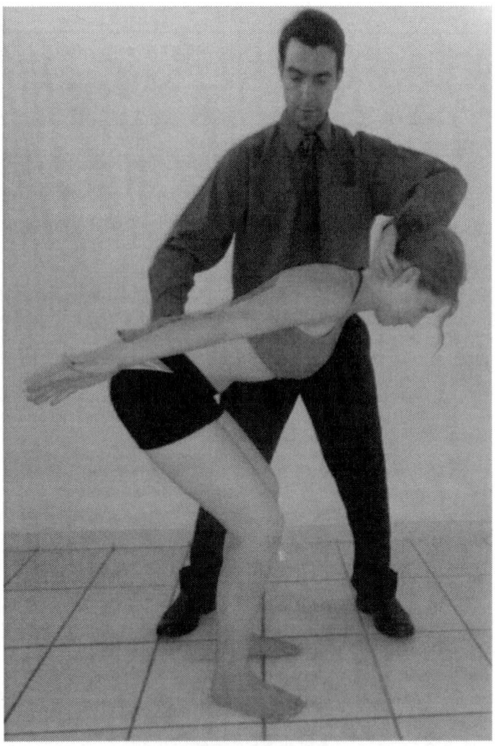

Figure 10. Shaktyasana - The shakty goddess pose.

Comparison of Chains

The most compromised chain is the one presenting the highest compensation in the test. This chain is the first to be treated.

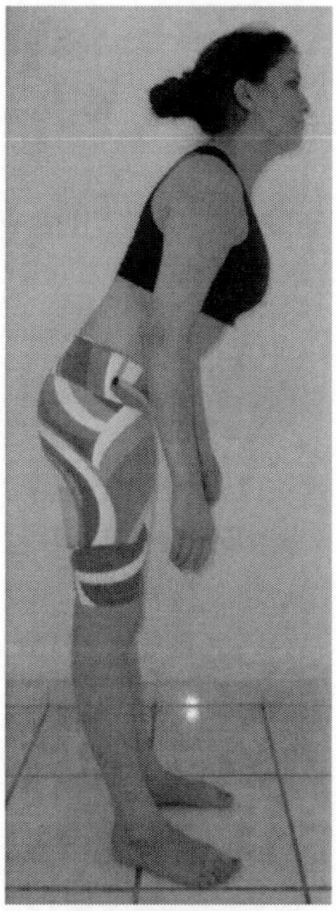

Figure 11. Compensation for the posterior chain test: opening the hip angle of flexion.

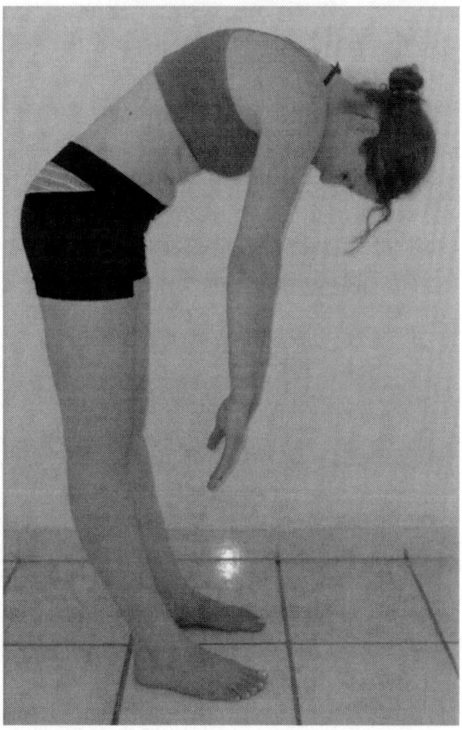

Figure 12. Compensation for the posterior chain test: opening the hip angle of flexion and opening of the ankle angle.

Examination of the Arm Chains

If the anterior chain is the most compromised, this test should be done in a standing (Figure 13) position, with rectification of the lumbar spine. If the posterior chain is the most compromised, the test should be performed seated, holding a small lordosis (Figure 14). After this positioning, it is essential to confirm that the thoracic curve is not flattened and to add adduction and depression of the scapula, external rotation of the arm, forearm supination and extension of the wrist and fingers.

After aligning the posture, it is necessary to observe the compensation generated by both abduction and then adduction of the upper limbs. If abduction generated more compensation than adduction of the upper limbs, it was given priority in terms of treatment. If adduction compensated more, adduction was the priority.

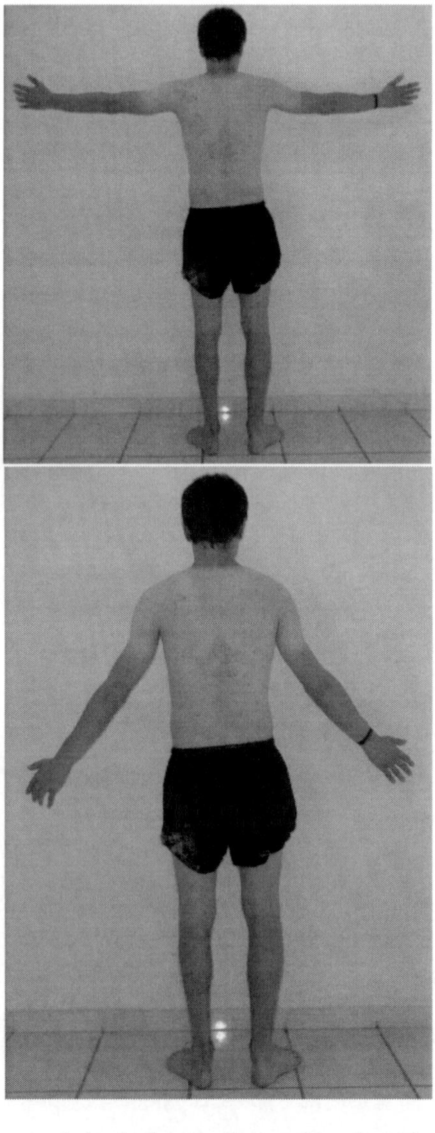

Figure 13. Examination of the arm chains in the standing position (anterior chain shortened).

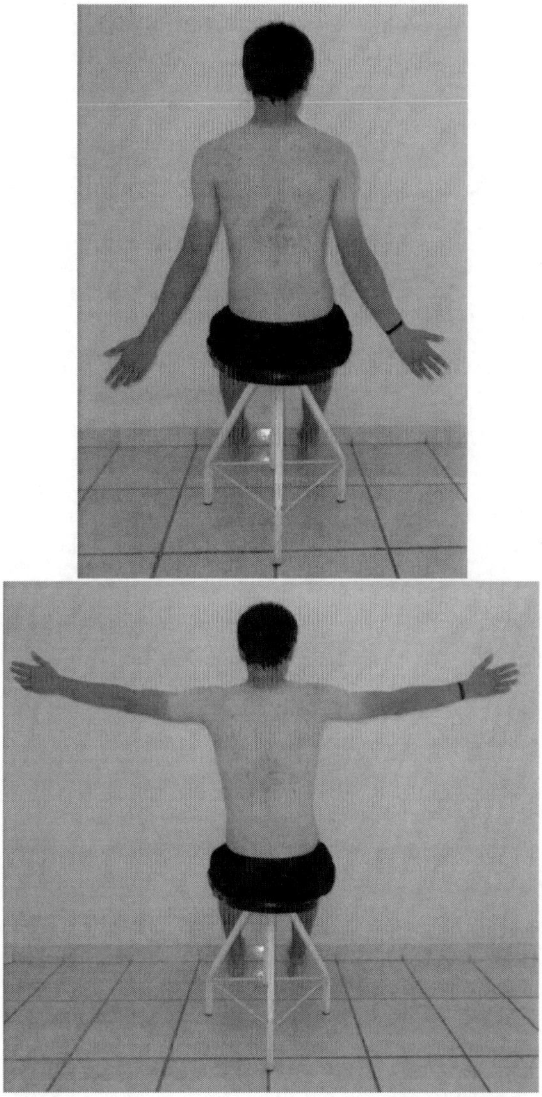

Figure 14. Examination of the arm chains in the seated position (posterior chain shortened).

Treatment

Treatment consists of two 15-minute postures. Before treatment, the patients were taught how to separate their breathing by region: apical; lower ribs and diaphragmatic breathing, in order to help the maintenance of posture. The selection of posture was based on the assessment described above. If there were more alterations in the posterior chain, two postures of the posterior chain were performed. If there were more alterations in the anterior chain, two postures of the *anterior chain* were performed. If the two chains exhibited similar alterations, treatment involved one posture of each chain. The therapist should choose a lie down posture in the following circumstances: 1 - more work was needed on breathing; 2 -

pain or disability prevented the patient from remaining standing or sitting; 3 - if the work was focused on the head, neck or upper limbs. Postures with load (sitting or standing) should be selected in the following cases: 1 - focus on the trunk or lower limbs; 2 – to provide a stronger stretch or improve muscle strength.

All postures must be performed actively by the patient, helping to promote proprioception and eccentric stretching. It is possible to perform at least two postures in one session. Sessions are conducted once or twice a week, depending on whether the problem to be treated is chronic or acute. Acute cases may involve less therapy time and a greater session frequency. The muscles evolved in each of the modified yoga positions are those found in the posterior and anterior muscular chains. [12, 28]

The following is a description of postures according to Rosário [12]:

Supta Baddha Konasana – Reclining Bound Angle Pose (Anterior Chain) (Figure 15)

- The patient is positioned in the supine decubitus, with the arms against the body;
- The patient puts the soles of the feet together;
- As a rule, the total external rotation of the femur should be sought. If the patient has excessive external rotation of the femur, a neutral position of femur rotation can be adopted;
- Neck traction while maintaining physiological neck lordosis;
- The tension point of the posture can be found by bringing the heels forward, extending the knees, leading to more difficulty in keeping the patient's lower back flat against the table.

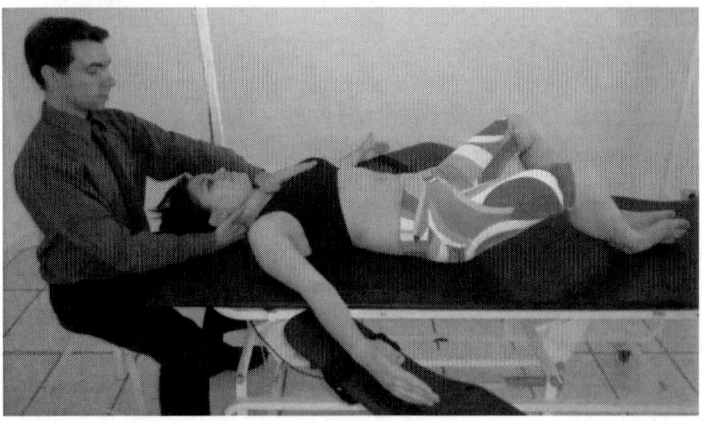

Figure 15. Supta baddha konasana – reclining bound angle pose.

Viparita Karani – Inverted Legs Pose (Posterior Chain) (Figure 16)

- The patient is positioned in the supine decubitus, with the arms against the body;
- The patient puts the soles of the feet together;
- As a rule, the total external rotation of the femur should be sought. If the patient has excessive external rotation of the femur, a neutral position of femur rotation can be adopted;

- Neck traction while maintaining physiological neck lordosis;
- The therapist flexes the patient's hip holding them by the heels. Alternatively, the therapist can use a support for the heels (Figure 16) in order to free the hands for treatment;
- The tension point of the posture can be found by bringing the heels forward, extending the knees, leading to more difficulty in keeping the patient's lower back flat against the table.

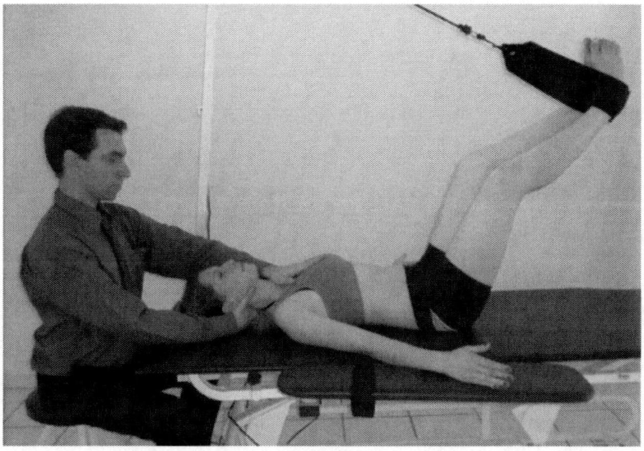

Figure 16. Viparita Karani – inverted legs pose.

The standing postures were not used in this study aiming a better control of the scientific variables.

Postures Evolution

During the 15 minutes-posture period, the therapist must seek to maintain the symmetry of the patient. When the patient maintains a posture for a certain time, muscular viscoelasticity decreases. As a result, the posture becomes easier to sustain. Thus, the difficulty of the posture must be gradually increased. The name of this process is posture evolution. In other words, evolution involves finding a new tension point each time the position starts to become easier. The following rules must be respected:

- At no time is it permissible to reverse the spinal physiological curves;
- Both anterior and posterior postures require gradual extension of the knees;
- The posture of the posterior chain requires a gradual increase in hip flexion.

Pain Assessment

Pain is a symptom that accompanies the majority of pathological conditions that require medical care transversely. Among the internationally validated scales for measuring the intensity of pain, the analog scale is one of the most utilized. [29-31] In the present study, a line, scored from 0 to 10, was shown to the subject, who verbally identified the degree of their pain, with zero indicating no pain and ten indicating the most intense pain possible.

Statistical Analysis

Besides the descriptive analysis, the Student's t-test was used, with the significance level set at $p = 0.05$. Paired Student's t-test was used when comparing the pain relief after the treatment with the pain values before the treatment in the same group. Student's t-test for two samples with equal variance was used when comparing the groups.

Results

Table 1 displays the number of times that each score (from zero to ten) was reported for all of the 126 spinal pain complaints of the PCT group, before and after treatment.

Table 1. Number of times each value of the analog scale was reported in the 126 complaints, before and after one session of muscular chain therapy

	Amount of occurrence of each value of the analog scale	
Value on the analog scale	Before treatment with PCT (n = 126)	After treatment with PCT (n = 126)
10	0	0
9	5	0
8	15	3
7	18	0
6	28	0
5	29	3
4	8	18
3	18	10
2	5	13
1	0	38
0	0	41

Before the PCT session, the greatest number of complaints was between the values 5 to 8. After the session, the majority of values were between 0 and 1, with a significant number of complaints also found at value 2, 3 and 4 of the scale. The no pain value (zero) increased from zero reports to forty one, from the total of one hundred twenty six complaints.

The mean of the 126 values before treatment was 5.55 with a standard deviation of 1.62. After treatment, these values dropped to a mean of 1.79 and a standard deviation of 1.79. The number of complaints with some improvement was 122, which is equal to 97% of cases. PCT had no effect in only 1 case (0.7%). In addition, there were also 3 cases (2.3%) that worsened. The difference between the before and after averages is 3.93. The significance of these data in the Student's t-test was $p = 0.00001$. Thus, it is possible to reject the equality between values before and after treatment. Table 2 displays the values for 50 complaints of the control group.

Table 2. Number of times each value of the analog scale was reported in the 109 complaints, before and after one session of turned off ultrasound (placebo)

	Amount of occurrence of each value of the analog scale	
Value on the analog scale	**Before treatment with placebo (n = 109)**	**After treatment with placebo (n = 109)**
10	0	1
9	5	3
8	12	9
7	14	8
6	23	15
5	18	16
4	15	18
3	20	25
2	2	12
1	0	0
0	0	2

Before the placebo session, the largest number of complaints was between the values 3 to 8. After the session, the majority of values were between 2 and 6. On the scale, the value of zero (no pain) increased from zero reports to one.

The mean of the 109 values before treatment was 5.42 with a standard deviation of 1.82. After treatment, these values dropped to a mean of 4.64 and a standard deviation of 2.07. The number of complaints with some improvement was 40, which is equal to 36.7% of cases. The placebo therapy had no effect in 64 cases (58.7%) and in 5 cases (4.5%) the pain worsened. The difference between the before and after averages is 0.78.The significance of this data in the Student's t-test was $p = 0.0001$.

Comparing the values before the treatment between groups with the Student t-test showed a p value of 0.9. Therefore, it is not possible to reject the equality between the groups before treatment. Comparing the difference before and after treatment between the groups with the Student's t-test, the value of p was 0.0001 (Table 3).

Table 3. Mean and standard deviation before and after treatment. Intergroup p value and p value between the groups

	Mean and standard deviation before treatment	Mean and standard deviation after treatment	Intergroup p value	p value between the groups
MCT Group	5.55 $\pm$ 1.79	1.62 $\pm$ 1.79	0.00001	0.0001
Control Group	5.42 $\pm$ 1.82	4.64 $\pm$ 2.07	0.0001	

Discussion

Both Groups showed significant relief of undiagnosed spinal pain. The placebo exhibited a p-value of 0.0001, whereas the value for yoga was 0.00001 after treatment. Comparing the groups, yoga was significantly better with a p value of 0.0001. The control group had a mean pain level of 5.42 before the treatment and 4.64 after (a difference of 0.78). The mean pain level in the yoga group was 5.55 before treatment and 1.79 after (a difference of 3.93). Treatment with yoga provided some sort of pain relief in 97% of cases, whereas this figure was 36.7% in the control group. In a previous study, [32] the same technique was applied for any type of musculoskeletal pain including the spine, but not restricted to this region, with similar results.

Although there are relatively few scientific studies approaching these yoga postures or similar techniques based on Meziérès stretching techniques [MST], they have produced results in treating various musculoskeletal conditions. [15-23, 33] The present study corroborates these previous experiments.

Canto et al. [16] studied the efficiency of a MST in individuals with lower back pain, in terms of the level of pain using visual analog pain at the time of the first and tenth treatment session. In total, 85.7% of the participants reported a decrease in the level of pain at the end of treatment and 77.1% of the subjects recorded a lower score on the Roland Morris questionnaire. The result of 85% is close to the 97% found in the present study, although the main difference is the number of sessions: while these authors used ten sessions, the present study used only one. This brings into question the number of sessions necessary to perform the treatment. Most MST therapists perform ten sessions as a basis for a treatment that provides results. [12] Moreover, since Canto et al. [16] did not assess patients before the eleventh session, but after the tenth, these researchers have assessed the pain with the acute effect of the last session and not with the chronic effect of the ten sessions alone. The present study also did not assess the effect of a single session after 24 or 48 hours. This form of data would be very interesting, in terms of finding a more effective treatment.

However, if MST or PCT treat pain through postural improvement, it is possible that these musculoskeletal pains will not return after full treatment. Moreira et al. [20] studied a group of five women aged between 20 and 30 years. The women were submitted to physical therapy to correct their posture and to reduce the pain caused by postural abnormalities. The patients were radiographed one week after therapeutic discharge. The postural improvement was evidenced by the retraction of the shoulder. Despite the small number of patients, the radiographic image is a high point of the study. Based on this study, it is possible to see that some of the complaints of musculoskeletal pain may be related to postural problems.

Another indication that postural correction may be the answer for the pain reduction is in the work of Rossi et al. [23] who obtained the postural improvements with just one treatment session using a similar technique. They assessed the effect of an application of the lying hip extension posture on 11 photographic postural variables. Of these variables, only 4 recorded significant improvements, and these four were all related to the head. The present study recorded pain reduction with one session, which can be related to that better posture achieved in the very first session. Rosário [24] obtained results in 3 of 6 variables with eight postural reeducation sessions, and there was no improvement in the shoulders and head. This highlights the fact that therapists place emphasis on postural correction, which may be more influential in a chosen segment. These studies revive the discussion, with a focus on posture and not on pain, about the time required for postural corrections and relief from the related pain. Rossi et al. [23] obtained the postural result with one posture and the present study recorded pain reduction with one session and two postures.

Marques et al. [19] assessed the effect of postural treatment on fibromyalgia. Twenty patients that had been diagnosed with fibromyalgia were treated for six sessions on average. Of the 20 patients, 18 reported some improvement, and 65% rated it as excellent or good, whereas 25% reported it as fair and only 10% reported no improvement. Although fibromyalgia is listed as a rheumatologic disease, these data are similar to the results of the present study.

Basso et al. [15] decreased the pain of 20 patients with temporomandibular disorders, using 10 muscular chain treatment sessions. In a further study, Gil et al. [17] reduced back pain in pregnant women in 8 weeks. Using analog scales, Heredia and Rodrigues [34] relieved the pain of patients with epidural fibrosis in post-operative lumbar disc herniation with 15 sessions.

Using a pressure platform (Tekscan-Matscan), Teodori et al. [35] undertook an interesting case study and noted changes in plantar pressure distribution and the location of the center of force in a subject with a history of right ankle sprain, with free bipedal support and with the eyes open. Asymmetry was found in the distribution of plantar pressure applied to the subject in one postural reeducation session, and was followed by an assessment of the pressure platform immediately after the intervention, and after 7, 14 and 30 days. The results clearly showed a recovery of symmetry, which continued for 7 days. After this period, there was a gradual recovery of asymmetry, although the initial values had not been attained after 30 days. Although a case study, if the data for this author is correct for the general population and most musculoskeletal pain is related to posture, one MCT session, as performed in the present study, would have the effect of pain resolution for seven days, and would remain in effect for 30 days or more in cases of chronic pain.

Rosário et al. [27] argue that this type of postural treatment technique does not act on posture simply by stretching, since a 15-minute posture provided similar results for hamstring flexibility as a 30-second hamstring stretch. Body awareness and the active maintenance of better joint positioning, reducing an existing subluxation, can exert their influence on postural adjustment and consequently solve related pain. [36] Whatever the reason for the effect, this study and previous studies have shown that MST causes a significant improvement in the efficiency of a musculoskeletal pain source. The relief of 94.4% of the complaints, with a complete absence of pain in 24 of the 71 complaints, demonstrated the efficiency of yoga modified postures when properly applied.

An interesting result in the present study was the success of the placebo therapy, which was inferior to the Yoga treatment but still provided some pain relief in 36.7% of cases. These results were similar to the ones found by Rosário et al. [32] who used the same method to treat any type of undiagnosed musculoskeletal pain and used a disconnected ultrasound as a placebo therapy, which had a 36% rate of success. These data also support the findings of Beecher [36] who suggested that the placebo effect occurs in 35% of the population. Placebo analgesia for post-operative dental pain was effective in 39% of cases. [37] Grelotti and Kaptchuk [38] discussed that the placebo effect comes from an emotional response and can be so strong that not only the patient feels better, but members of the family can also think the treatment is working, provoking a whole social effect. These results corroborate the findings of other authors who suggested that the placebo effect occurs in 20% to 40% of the cases. [39]

PCT was shown to reduce musculoskeletal pain in young adult patients immediately post intervention. However, our study did not determine short term or the long term effects of PCT. Further studies are required to verify the short and long-term effect evaluation and optimal treatment duration for postural improvement and the decrease of pain. Objectives measures are also recommended for future works.

Conclusion

This chapter showed that the PCT, using modified yoga positions, was shown to reduce musculoskeletal pain in patients immediately after the intervention, probably because of a postural improvement. However, it did not determine the short term or long term effects of just one intervention with yoga. Further studies are required to understand these effects and the disorders that could be treated effectively by this method, as well as those that prove unresponsive. Moreover, identifying the optimal timeframe and frequency of application for each disorder should be determined.

References

[1] Rosário JLP, Marques AP, Maluf SA. Aspectos clínicos do alongamento: uma revisão de literatura. *Revista Brasileira de Fisioterapia*. 2004;8:83-88.

[2] Kendall FP, McCreary EK, Provance PG. Músculos – Provas e Funções. São Paulo: Manole; 2010.

[3] Novak CB, Mackinnon SE Repetitive use and static postures: a source of nerve compression and pain. *J Hand Ther*. 1997;10:151-9.

[4] Lee D. Principles and practices of muscle energy and functional techniques. In: Grieve GP (ed.) Modern manual therapy of the vertebral column. New York: Churchill Livingstone; 1986. p. 279-301.

[5] Kappler RE Postural balance and motion patterns. *J Am Osteopatic Assoc*. 1982; 81:69-77.

[6] Orchard J, Marsden J, Lord S, Garlick D Preseason hamstring muscle weakness associated with hamstring muscle injury in Australian footballers. *Am J Sports Medicine*. 1997; 25:81-5.

[7] Bennell K, Wajswelner H, Lew P, Schall-Riaucour A, Leslie S, Plant D, et al. Isokinetic strength testing does not predict hamstring injury in Australian Rules footballers. *Brit J Sports Medicine*. 1998; 32:309-14.

[8] Mirtz TA, Morgan L, Wyatt LH, Greene L. An epidemiological examination of the subluxation construct using Hill's criteria of causation. *Chiropr Osteopat*. 2009; 2;17:13.

[9] Rosário JLP, Marques AP, Maluf AS. Aspectos Clínicos do alongamento: uma revisão de literatura. *Revista Brasileira de Fisioterapia*. 2004; 8:83-8.

[10] Moreira CMC, Soares DRL. Análise da efetividade da reeducação postural global na protusão do ombro após a alta terapêutica. *Fisioter Mov*. 2007;20:93-9.

[11] Teodori RM, Negri JR, Cruz MC, Marques AP Reeducação postural global: uma revisão da literatura. *Rev Bras Fisioter*. 2011;15: 185-9.

[12] Rosário JLP. Manual Prático de Reeducação Postural: o que você precisa saber para um tratamento eficiente. São Paulo: Ed. Baraúna; 2011.

[13] Bertherat T, Bernstein C. O corpo tem suas razões – Antiginástica e a consciência de si. São Paulo: Martins Fontes; 1987.

[14] Hoppenfeld S. Propedêutica ortopédica coluna e extremidades. São Paulo: Atheneu;1999.

[15] Basso D, Corrêa E, Silva AM. Efeito da reeducação postural global no alinhamento corporal e nas condições clínicas de indivíduos com disfunção temporomandibular associada a desvios posturais. *Fisioter Pesqui*. 2010;17: 63-8.

[16] Canto CREM, Oliveira LF, Gobbi FCM, Theodoro MN. Estudo da eficácia do metódo de reeduçação postural global em indivíduos com dor lombar com relação a dor e incapacidade funcional. *Ter Man*. 2010;38: 292-7.

[17] Gil VFB, Osis MJD, Faúndes A Lombalgia durante a gestação: eficácia do tratamento com Reeducação Postural Global (MCT). *Fisioter Pesqui*. 2011;18: 164-70.

[18] Luz, GCP, Cheik NC, Ferreira F, Pereira PAC, Vidal JS, Affonso F, Baraúna MA. Tratamento da lombalgia através do dispositivo lombo abdominal e da reeducação postural global. *Ter Man*. 2008; 6:287-92.

[19] Marques AP, Mendonça LLF, Cossermelli W. Alongamento muscular em pacientes com fibromialgia a partir de um trabalho de reeducaçao postural global (RPG). *Rev Bras Reumatol*. 1994;34: 232-4.

[20] Moreira CMC, Soares DRL. Análise da efetividade da reeducação postural global na protusão do ombro após a alta terapêutica. *Fisioter Mov*. 2007;20: 93-9.

[21] Moreno MA, Catai AM, Teodori RM, Borges BLA, Cesar MC, Silva E, et al. Efeito de um programa de alongamento muscular pelo método de Reeducação Postural Global sobre a força muscular respiratória e a mobilidade toracoabdominal de homens jovens sedentários. *J Bras Pneumol*. 2007; 33: 679-86.

[22] Moreno MA, Catai AM, Teodori, RM, Borges BLA, Zuttin RS, Silva E, et al. Adaptações do sistema respiratório referentes à função pulmonar em resposta a um programa de alongamento muscular pelo método de reeducação postural global. *Fisioter Pesqui*. 2009;16: 11-5.

[23] Rossi LP, Brandalize M, Gomes ARS. Efeito agudo da técnica de reeducação postural global na postura de mulheres com encurtamento da cadeia muscular anterior. *Fisioter Mov*. 2011;24: 255-63.

[24] Rosário JLP, Nakashima IY, Rizopoulos K, Kostopoulos D, Marques AP. Improving posture: Comparing segmental stretch and muscular chains therapy. *Clin Chiropractic*. 2012; 15: 121-8.

[25] Myers T. Anatomy Trains. 2nd ed. Edinburgh: Churchill Livingstone / Elsevier; 2009.

[26] Marques AP. Cadeias musculares – um programa para ensinar avaliação fisioterapêutica global. Manole, São Paulo; 2005.

[27] Rosário JLP, Sousa A, Cabral CMN, João SMA, Marques AP. Reeducação postural global e alongamento estático segmentar na melhora da flexibilidade, força muscular e amplitude de movimento: um estudo comparativo. *Fisioter Pesqui*. 2008;15: 12-18.

[28] Karminoff L. Yoga anatomy. Champaign IL: Human Kinetics; 2007.
[29] Gracely RH, Price DD, Roberts WJ, Bennett GJ. Quantitative sensory testing in patients with CRPS-I and -II. In Janig W, Stanton-Hicks M. (Eds.). Reflex sympathetic dystrophy – A reappraisal. Seattle: IASP Press;1996. p.151-172.
[30] Aicher B, Peil H, Peil B, Diener HC. Pain measurement: Visual analogue scale (VAS) and verbal rating scale (VRS) in clinical trials with OTC analgesics in headache. *Cephalalgia*. 2012;32:185-97.
[31] Bailey B, Gravel J, Daoust R. Reliability of the visual analog scale in children with acute pain in the emergency department. *Pain*. 2012;153:839-42.
[32] Rosário JLP, Orcesi LS, Kobayashi FN, Aun AN, Diolindo Assumpção IT, Blasioli GJ, Hanada ES. The immediate effects of modified Yoga positions on musculoskeletal pain relief. *J Bodyw Mov Ther*. 2013;17:469-74.
[33] Fozzatti MCM, Palma P, Herrmann V, Dambros M. Impacto da reeducação postural global no tratamento da incontinência urinária de esforço feminina. *Rev Assoc Med Bras*. 2008; 54: 17-22.
[34] Heredia EP, Rodrigues FF. O tratamento de pacientes com fibrose epidural pela reeducação postural global. *Rev Bras Neurol*. 2008;44: 19-26.
[35] Teodori RM, Guirro ECO, Santos RM. Distribuição da pressão plantar e localização do centro de força após intervenção pelo método de reeducação postural global: um estudo de caso. *Fisioter Mov*. 2005;18: 27-35.
[36] Beecher HK. The powerful placebo. *J Am Med Assoc*. 1955; 159: 1602-6.
[37] Levine JD, Gordon NC, Bornstein JC, Fields HL. Role of pain in placebo analgesia. *Proc Natl Acad Sci USA*. 1979; 76: 3528-31.
[38] Grelotti DJ, Kaptchuk TJ. Placebo by proxy. *BMJ*. 2011; 11:343-5.
[39] Verhulst J, Kramer D, Swann AC, Hale-Richlen B, Beahrs J. The medical alliance: from placebo response to alliance effect. *J Nerv Ment Dis*. 2013; 201:546-52.

In: Posture: Types, Exercises and Health Effects
Editor: Sarah A. Curran

ISBN: 978-1-63117-252-6

Chapter 11

Yoga Postures and Colon Cleanse

Vijaypal Arya, * ***MD***
Cornell University, New York, US
North Shore LIJ Health System, New York, US
Endoscopy Unit, WHMC, Brooklyn, New York, US

Abstract

This chapter presents a novel approach to colon cleanse. No harsh laxative chemicals such as phosphates, sulphates, or citrates are used. The process emphasizes the basic tenet of yoga (unifying mind, body and spirit). The innovative aspect involves a bolus drinking of lukewarm saline in conjunction with sequential posture changes. In addition, deep breathing, relaxation and meditation helps in bowel cleansing by parasympathetic activation. This yoga based purification technique can be learned by following a DVD. On an average, it takes less than two hours to complete the process and is ideal for healthy men and women who have an open mind for yoga and are fit enough to do a little exercise.

Keywords: Posture, saline, colon cleanse, yoga

Introduction

Physicians have been prescribing laxatives for centuries, but the need to cleanse the colon was first recognized by the radiologist circa 1895 after the invention of radiographs and the use of barium enema to study the large bowel. [1] Subsequently, surgeons realized that if the colon is clean before abdominal surgeries, the incidence of post operative infections can be reduced. [2, 3] Initially, a large volume (9-12 liters) of normal saline/balanced electrolyte solutions were used to cleanse the bowel. [4, 5] Patients were either asked to drink at a rate of

* Author contact: varyamd@outlook.com

25-30 ml/minute (usual sipping rate) or solutions were infused via naso-gastric tube for about 4-6 hours to complete the cleansing process. In 1957, the introduction of the first fiber optic endoscope revolutionized the medical field and a new subspecialty emerged – *gastrointestinal endoscopy.* [6] For a meticulous examination of the colonic mucosa, the lumen should be completely devoid of waste material, particularly if the aim is to recognize and remove abnormalities as small as a pin head (5-6 mm). These are called polyps and are of many different histologic subtypes. The adenoma type is pre-cancerous and early removal can prevent colon cancer. [7] Colon cancer is the third most common cancer in United States and is the second leading cause of cancer deaths. [8] Screening and surveillance colonoscopy and polypectomy has already reduced the incidence of colon cancer deaths, the poor bowel prep (about 10-15%) remains a significant limiting factor in the success of this life saving procedure. [9] Many patients dread the task of cleaning more than the procedure itself. The large volume (4L) and unpalatable taste of the most commonly used polyethylene glycol (PEG) based preps are a popular deterrent among patients. [10] An ideal preparation should be effective, safe, palatable, fast and impose few dietary restrictions on the patients. While no prep regimen can be ideal for all groups of patients, a novel approach has been explored, which is based on yogic principles. [11, 12] A kriya (sanskrit word for process) known as shankh prakshalana has been standardized as “shudh” (pronounced “should” and in sanskrit means purity). This process involves drinking lukewarm saline in bolus form (large volumes, 240ml, taken intermittently) alternating with five yoga asanas or postures. This represents a potential time saving alternative, which is economical as well.

The knowledge presented in this chapter highlights tremendous medical value of a centuries-old practice. The process has been standardized so it can be used easily and without supervision by a yoga teacher. The basic mechanism of action has been delineated.

Scope of Complementary and Alternative Medicine (CAM)

More than one third of American adults use some form of CAM. Approximately 20 million individuals practice yoga, spending $10.3 billion a year. [13] Credible scientific research is published in peer reviewed journals and added to the literature every day. The medical value of CAM therapies is being recognized and is starting to be practiced. The common CAM therapies include acupuncture, tai chi, recki, and yoga. The word "yoga" comes from sanskrit, the most ancient Indian language and has two meanings. The first is union (union of spirit, mind and body). The root of union is "yugir". The second meaning is smadhi (The absolute knowledge, the highest state of mind and the final realization). The root of samadhi is yuja.

The average person understands yoga as simply a physical practice – this is a huge misconception. Yoga actually has three components: s*piritual, mental, and physical*. The goal is to unify spirit, mind, and body (Figure1). In its most original form, described by Rishi Patanjali in the yoga darshan, yoga has eight limbs and is called ashtanga ("asht-" meaning eight, and "-anga" meaning limbs) yoga. The eight limbs are: *yama* (duties towards others, of which there are five), *niyama* (duties towards oneself, of which there are also five), *asana* (postures), *pranayama* (breathing exercises), *pratyahara* (introspection), *dharana* (concentration), *dhyana* (meditation), and *samadhi* (state of super-consciousness). The last three limbs describe advanced meditative practices through which the practitioner can achieve autonomic nervous system control.

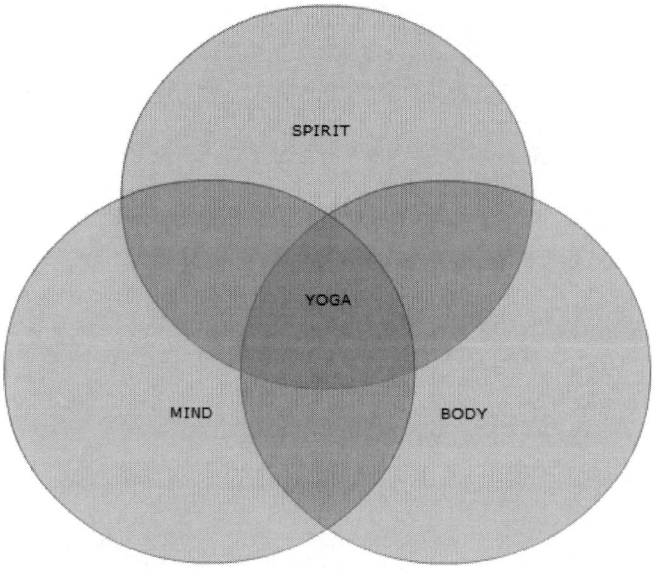

Figure 1. The union of spirit, mind and body is yoga.

The first niyama is *saucha* (pronounced shau-cha), which means cleanliness of mind and body. Shankh-prakshalana has been defined in yoga literature as a means of cleansing the digestive tract. In sanskrit, shankh means "conch." The sea shell, with its convoluted chambers, is a metaphor for the gastrointestinal tract. Prakshalana means "to clean." In an effort to make this millennia-old method more widely known, Arya et al. [11] have simplified shankh-prakshalana to shudh. The average time to complete the cleanse process is less than two hours, as compared to four to six hours with conventional methods.

Shudh Process

The first pre-requisite to practice yogic processes is willingness to perform. It is the authors' advice to health practitioners to experience the shudh process themselves before prescribing it to the patients. For someone with no prior knowledge and experience the only way is to take a leap of faith and try. To practice shudh one should either watch the DVD (www.shudhinc.com) and learn the poses or read up the instructions and practice the poses in advance. All the poses are simple to practice except the fifth pose (demands squatting) and can be modified as explained. The best time to perform the process is early morning after a good night's sleep. The body is well rested and all the nutrients have passed from the small bowel into the large bowel. The morning time is also the time for increased colonic motility and natural bowel movement. It's advised to eat a light dinner the day before the process. If the purpose of doing the shudh-process is cleansing the colon before colonoscopy, one should eat light lunch by 2.00 pm the day before and then stay on liquids for the rest of the day until going to sleep. Clear broth, gelatin, apple juice, and white grape juice is fine, but nothing red and nothing with pulp. It is also advised to avoid eating or drinking anything in the morning. This dietary restriction is necessary to achieve high quality colon cleansing, so that the doctor performing the colonoscopy will be able to perform a high quality examination. To begin, wear comfortable and loosely fitting clothing and choose a quiet environment and firm surface. A dedicated bathroom should be available for use. All distractions should be avoided i.e., turn off phones and beepers, relax and practice mindfulness. To prepare the solution dissolve one sachet of pre-packaged 9 gm of NaCl (www.shudhinc.com) in one liter of lukewarm water. A twist of lemon/lime can be added according to palatability preferences.

1. The actual shudh process includes – drinking lukewarm saline in bolus forms (rather than sipping) and then perform yoga poses (Table 1):

- Drink one or two 8-ounce glasses of the solution. Do the upward stretch and the side stretch;
- Drink one or two 8-ounce glasses. Do the twist stretch and the push-up stretch;
- Drink one or two 8-ounce glasses. Do the squat stretch;
- Repeat the cycle.

The rapid drinking is what brings the rapid results. This works like a flush while the stretching exercise works like a pump until it all goes through and one comes out clean. An average person needs to drink 2-3 liters of solution and completes 2-3 cycles of yoga poses to complete the process. Sometimes fullness in the stomach and feeling of nausea precludes the required drinking. In this situation, skip the drinking; continue the yoga poses until ready to drink again.

One can expect to have first bowel movement between 20 and 40 minutes. The stool comes out in a rush; hence slightest urge to defecate should not be ignored. It's advised to interrupt the process to use the toilet and avoid straining. The process should be continued where left off. Individual results vary and sometimes a delayed response, up to few hours has been noticed. The average time to complete the process is about 2 hours. At this point the excreta would look like urine. Stop the process and take rest in supine position also called shav-asana for 20-30 minutes.

Upward Stretch - Palm Tree Pose - Tad-asana (Taad'as'ana)

"Tad" pronounced – taad, means "palm tree" and "asana" means posture. This posture resembles a palm tree. To begin, stand up straight, interlace fingers of both hands and put them on your head with the palm facing up. While inhaling stretch the body up and stand on your toes, like you're trying to reach the sky. Keep your knees straight. Look straight in the infinity. Wait for 5 seconds, while exhaling come to the ready position. To practice mindfulness and achieve concentration, try to feel your breath as it enters through your nostrils and imagine that all positive energy is coming inside you. At the same time while exhaling imagine that all toxins are leaving your body. Do not occupy your mind with any thoughts but to achieve cleansing of your body. Relax your shoulders when you come down. Repeat this pose ten times.

Side Stretch - Oblique Palm Tree Pose - Tiryak Tadasana (Tir'yak'taad'as'ana)

To get ready, stand straight, move your feet 1 – 2 feet apart, interlace your fingers, stretch arms upwards, palms facing up. While inhaling, bend to the right side; try to bend directly to the side, not forward or backward. Try relaxing your neck and letting your head flop over when you bend. After waiting for five seconds, exhale slowly and come back to the ready position. Take a resting break and then bend to the left side. Repeat this stretch five times on right and five times on left.

Twist Stretch - Waist Twisting Pose - Katichakrasana (Khat - Ee'chak'ra'sana)

Start off with pulling feet apart and arms stretched out horizontally, taking up the position of the famous DaVinci drawing. Put the left hand on your right shoulder and the right hand on the small of your left back, the palm facing out. Inhale and start with the twist while looking back at your left heel. Wait for five seconds, then exhale and come back to the ready position. Now repeat the pose on other side. Repeat this stretch five times on right and five times on left.

Turtle Stretch - Twisting Cobra - Tiryak Bhujangasana (Tir'yak'bhuj'ang'asana)

Lie down on your stomach with the feet apart, like doing a push up. Push your upper body off the floor leaving the hips on the ground. While inhaling push yourself up as you lift your chest, and twist your upper body to the right side, try to look at your left heel. Wait for five seconds the exhale and come to ready position. Repeat this stretch five times on right and five times on left.

Squat Stretch - Abdominal Stretch Pose - Udarakarshanasana (Ud'ara'kar'asana)

Grab onto your knees with your feet apart and gently lower yourself into a squat. Stand on the balls of your feet if you can. This is the ready position. Inhale and try to touch your left knee to the right toes. Rotate your rib cage around the spine, twisting your torso around looking up and over the shoulder. Exhale as you come back to center. Your hands should help with balance and steering your knees around. Repeat the pose on left side. You can hold on to a chair for support and squat as much as you can and just twist. Repeat this stretch five times on right and five times on left.

Precautions

The process should be practiced by healthy people only. According to the yogic literature the process can be done with plane lukewarm water in case of salt sensitivity and contraindications to salt intake. Anybody with poor heart, liver and kidney function are specifically at risk for salt and fluid overload and should not attempt this process. At the same time inability to perform poses due to arthritis and joint diseases and pregnant females should not practice this process. This also should not be considered as a treatment for severe constipation. The process should not be practiced more than once in 2-3 months.

Adverse Effects

Failure of the process meaning no bowel movements after drinking 2-3 liters of lukewarm saline has been experienced. This will also lead to benign swelling of the body resolving in few days. Sometimes increased urination has been reported instead of defecation. If feelings of nausea, dizziness, severe bloating, abdominal distension and pain persist or if bowel movement is not initiated after consuming approximately two liters of solution, stop the process. It is advised not to consume more than four liters of salt solution at a time.

Benefits - Detoxification

Modern medicine believes that colon / large intestine is a self-cleansing organ, hence it does not need to be cleaned. The users of shudh, anecdotally have reported, the feeling of rejuvenation and being full of energy after the process. This is in contrast to feeling weak after using other bowel cleansing methods and harsh laxatives. Further research is needed to prove this point. For those who believe in colon cleansing, shudh is the most natural, economic and fast alternative. This can be learned by watching a DVD in the comfort of your home, without the worry of inappropriate social constraints, like embarrassment from rectal instrumentation during colonics.

Constipation is a very common problem worldwide. About 15% of Americans (60 million) suffer from this problem. [14] If left untreated chronic constipation can lead to complications like hemorrhoids, anal fissures, solitary rectal ulcers, and overflow incontinence. According to a recent report more than $700 million are spent on laxatives yearly by Americans alone. Although no study has been done using this process to treat constipation, intuitively the yogic cleansing process can be a very economical, simple and faster alternative to treat mild constipation.

Science of Shudh Process: Millennia Old Process and Modern Physiological Principles

Other than mindfulness, shudh process has two components: A - bolus (8oz-16oz in 1-2 minutes) drinking of lukewarm normal saline and B - yoga postures with deep breathing. According to yoga philosophy water should be taken in bolus form especially in the morning to generate an urge to defecate. The distension of duodenum initiates the gastro-colic reflex, well recognized in modern physiology. [15] The signals from the duodenum reach to the right colon, causing movement of content from right to left instantaneously, leading to defecation. Thus another concept of modern physiology plays an important role in bolus drinking is first order kinetics. [16] The isotonic saline empties from stomach based on the first order kinetics- 50% of volume empties out in 8-18 minutes, meaning higher volume will empty out by bolus drinking as compare to sipping. The dumping of large volume of isotonic saline in intestine rushes through causing minimal time for absorption and no significant electrolyte imbalance.

The conventional, PEG (polyethylene glycol) based colon preps are advised to drink chilled partly to mask the unpalatable taste. Animal studies have shown that warmer liquids cause esophageal and gastric relaxation, suggesting that practice of drinking lukewarm saline might be beneficial. [17, 18] The common practice also dictates use of warmer temperature for cleansing.

Role of Yoga Postures

Application of different body positions is well established in modern medicine. Some of the well known positions are supine, prone, left and right lateral, Trendelenburg, reverse Trendelenburg, and lithotomy. All these positions have specific use in clinical medicine. Significant effect of body position on gastric emptying has been noticed on radio-nuclide studies. [19] Simply lying on right side improves gastric emptying. [20] Position change together with gravity changes the gastric configuration, leading to intra gastric re-distribution and faster gastric emptying. [21, 22]

During the first pose, stretching of the body upwards (with deep breathing causing diaphragmatic movements) results in change in gastric configuration. Arm elevation is known to increase the tidal volume. [23]

The first pose resulting in linear acceleration of vestibular fluid of semicircular canals increases parasympathetic tone. [24] The second, third and fourth poses will also lead to the same effect. The twisting of spine will be associated with compression of abdominal organs especially intestinal tract. Additionally, the effects on the vestibular system also might play a role in faster gastric emptying. During, the fifth pose (squat stretch) – squatting is known to increase intra-abdominal pressure and straightening of anorectal angle which facilitate defecation. [25]

Table 1. Shudh postures and body effects

Follow Steps 1 to 8 and keep repeating these steps until you have clear bowel movement.							
Step 1	Drink 1-2 glasses, 8 Oz (240ml) each, of the prepared solution. Drink continuously rather than sipping.						
	Get Ready	**Inhale**	**Exhale**	**Inhale**	**Repeat**	**Movement/Muscles Involved**	**Body effects**
Step 2 Upward Stretch	 **Hands**: Interlace fingers, hands on head, palm facing up **Feet**: Feet together, stand erect.	 **Hands**: Stretch upwards. **Feet**: Stand on toes stretch up.	Return to ready position		The Upward Stretch ten times	Arm elevation (SITS group, Triceps brachii, Anconeus) Toe extension (extensor digitorum longus, extensor hallucis longus) Spine: stretch	Increased tidal volume Increased intra-gastric pressure, Increased trans-diaphrgmatic pressure
Step 3 Side Stretch	 **Hands**: Interlace fingers, stretch arms upwards, Palms facing up. **Feet**: 1-2 feet apart	 **Hands**: Bend both arms to right. **Feet**: 1-2 feet apart	Return to ready position	 **Hands**: Bend both arms to left. **Feet**: 1-2 feet apart	The Side Stretch five times on right and five times on left.	Lateral flexion of spine: Erector Spinae, External oblique, Internal oblique , Lattismus dorsi, Psoas major	Same as above
Step 4	Drink 1-2 glasses, 8 Oz (240ml) each, of the prepared solution. Drink continuously rather than sipping.						
Step 5 Twist Stretch	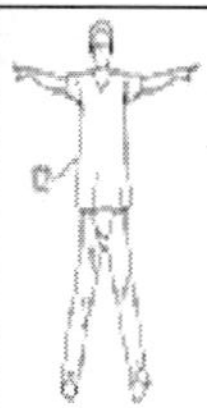 **Hands**: Stretch arms, palm facing down wards. **Feet**: 1-2 feet apart	 **Hands**: Left hand on right shoulder and right hand on small of back, palm facing out. Rotate your upper body towards the right. Try to see your left heel. **Feet**: 1-2 feet apart	Return to ready position	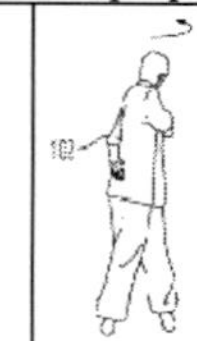 **Hands**: Right hand on left shoulder and left hand on small of back, Palm facing out. Rotate your upper body towards the left. Try to see your right heel. **Feet**: 1-2 feet apart.	The Twist Stretch five times on right and five times on left.	Spine: twist Erector Spinae, external oblique, internal oblique	Same as above

Follow Steps 1 to 8 and keep repeating these steps until you have clear bowel movement.							
Step 6 Push up Stretch	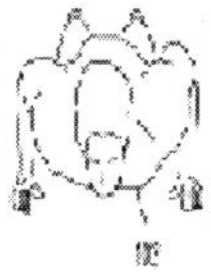 **Hands**: Lie down on stomach, raise your body. **Feet**: 1-2 feet apart	 **Hands**: Twist towards the right turning your head to the right until you see the heels. **Feet**: 1-2 feet apart	Return to ready position	 **Hands**: Twist towards the left turning your head to the left until you see the heels. **Feet**: 1-2 feet apart	The Push up Stretch five times on right and five times on left.	Spine extension and twist: Iliocostalis lumborum, Iliocostalis thoracis interspinalis, Intertransversalis, Longismus thoracis Arm extension:	Same as above
Step 7	Drink 1-2 glasses, 8 Oz (240ml) each, of the prepared solution. Drink continuously rather than sipping.						
Step 8 Squatting Stretch	 **Feet**: Squat on 1 feet foot apart. **Hands**: Palm on knees	 **Feet**: Touch left knee to right toe **Hands**: Palm on knees	Return to ready position.	 **Feet**: Touch right knee to left toe **Hands**: Palm on knees	The Squatting Stretch five times on right and five times on left.	Spine: Thoracic rotation Hip flexion: iliopsoas,sartorius,rectus femoris Hip adduction :adductor Magnus, brevis, longus, minimus, pectineus, gracilis, obturator externus Squatting: gluteus Maximus, quadriceps, hamstrings, Levator ani	Squatting posture is known to straighten the anorectal angle, which facilitates defecation.

At any point if you have an urge to defecate, go to the toilet but do not strain.

Role of Exercise on Gastrointestinal Motility

Acute aerobic exercise decreases colonic phasic motor activity causing less resistance to colonic flow. [26] The post exercise effect is increased amplitude of propagated waves leading to enhance propulsion of colonic contents. Simple walking has been shown to improve colonic preparation. [27] The exercise component of yoga postures are although mild intensity (metabolic equivalent value of three) should have a complementary role.

Role of Deep Breathing

Deep breathing is the most potent natural action to stimulate parasympathetic tone. [28] Deep breathing stimulates pulmonary stretch reflex, with the final effect of bradycardia, relaxation of intestinal sphincters and increased intestinal motility. [29]

Role of Mindfulness

Psychological factors (stress and relaxation) are well known to influence gut motility. [30] The role of turning inward - mindfulness, avoiding routine distractions -is difficult to measure. The process is performed in a calm environment. The postures are motivated by deep, rhythmic inhalation and exhalation with an aim to achieve relaxation of mind and body. Mind - body interaction has been identified by modern biomedical research. [31]

Conclusion

Shudh is a modern version of shankh prakshalana, a yogic purification technique. A new mechanism of colon cleansing has been explored. The process cleanses the colon naturally using only lukewarm saline water and a series of five yoga poses. Instead of chemical action the body's own reflexes do the work. The shudh process is ideal for healthy men and women who have open mind for yoga and are fit enough to do a little exercise.

References

[1] Rooney A. History of medicine. New York: Rosen Pub Group; 2012.

[2] Ellis H. The first successful elective laparotomy. *J Perioperative Pract.* 2008;18:211.

[3] Hewitt J, Reeve J, Rigby J Cox JA. Whole-gut irrigation in preparation for large-bowl surgery. *Lancet.* 1973;3:37-40.

[4] Levy AG, Benson JW, Hewlett EL, Herdt JR, Doppman JL, Gordon RS Jr. Saline lavage: A rapid, effective, and acceptable method for cleansing the gastrointestinal tract. *Gastroenterology.* 1976;70:157-61.

[5] Postuma R. Whole bowel irrigation in pediatric patients. *J Pediatric Surgery.* 1982;17:350-52.
[6] Irvin Modlin, A brief history of endoscopy. Milan: Multi Med; 2000.
[7] Winawer SJ, Zauber AG, Ho MN, O'Brien MJ, Gottlieb LS, Sternberg SS, et al. Prevention of colorectal cancer by colonoscopic polypectomy. The National Polyp Study Workgroup. *New England J Medicine.* 1993; 329:1977-81.
[8] Cummings LC. Colorectal cancer screening: update 2011. *Seminar in Oncology.* 2011: 38:483-489.
[9] Belsey J, Epstein O, Heresbach D. Systematic review: oral bowel preparation for colonoscopy. *Aliment Pharmacol Ther.* 2007;15; 25:373-84.
[10] Harewood GC, Wiersema MJ, Melton LJ. A prospective, controlled assessment of factors influencing acceptance of screening colonoscopy. *Am J Gastroenterol.* 2002;97:3186-94.
[11] Arya V, Gupta K, Arya S. Efficacy of bolus kukewarm saline and yoga postures as colonoscopy preparation: A pilot study. *J Altern Complementary Med.* 2010; 16:1269-77.
[12] Arya V, Gupta K, Valluri A, Arya SV, Lesser ML. Rapid colonoscopy preparation using bolus lukewarm saline combined with sequential posture changes: A randomized controlled trial. *Digestive Diseases Sci.* 2013;58:2156-66.
[13] Yoga Journal. "Yoga journal releases 2012 yoga in America market study" http://www.prnewswire.com/news-releases/yoga-journal-releases-2012-yoga-in-america-market-study-182263901.html. 5th December 2012. Retrieved on: 20th September 2013.
[14] Consumer health information corporation. http://www.consumer-health.com/services/LaxativesProceedwithCaution.php" Retrieved on: 29th October 2013.
[15] Tansy MF, Kendall FM. Experimental and clinical aspects of gastrocolic reflexes. *Am J Dig Dis.* 1973; 18:521-31.
[16] Hunt JN, Macdonald I. The influence of volume on gastric emptying. *J Physiol.* 1954;126:459-74.
[17] Bateman DN. Effects of meal temperature and volume on the emptying of liquid from the human stomach. *J Physiol.* 1982;331:461-67.
[18] McArthur KE, Feldman M. Intragastric temperature, gastric acid secretion, gastrin release, and gastric emptying following ingestion of hot, warm, or cold coffee in humans. *Gastroenterology.* 1984;90:1540.
[19] Moore JG, Datz FL, Christian PE, Greenberg E, Alazraki N. Effect of body posture on radionuclide measurements of gastric emptying. *Digestive Diseases Sci.* 1988; 33:1592-95.
[20] Burn-Murdoch R, Fisher MA, Hunt JH. Does lying on the right side increase the rate of gastric emptying? *J Physiol.* 1980;302:395-98.
[21] Amidon GL, DeBrincat GA, Najib N. Effects of gravity on gastric emptying, intestinal transit, and drug absorption. *J Clinical Pharmacology.* 1991; 31:968-73.
[22] Anvari M, Horowitz M, Fraser R, Maddox A, Myers J, Dent J, Jamieson GG. Effects of posture on gastric emptying of nonnutrient liquids and antropyloroduodenal motility. *Am Physiological Society Am J Physiol.* 1995;268:G868-71.
[23] Couser JI Jr, Martinez FJ, Celli BR. Respiratory response and ventilatory muscle recruitment during arm elevation in normal subjects. *Chest.* 1992;101:336-40.

[24] Carter JR, Ray CA. Sympathetic responses to vestibular activation in humans. *Am J Physiol Regul Integr Comp Physiol.* 2008;294:R681-8.

[25] Bassotti G, Germani U, Morelli A. Human colonic motility: physiological aspects. *Int J Colorectal Dis.* 1995;10:173-80.

[26] Rao SC, Beaty J, Chamberlain M, Lambert PG, Gisolfi C. Effects of acute graded exercise on human colonic motility. *Am Physiological Society.* 1999;276: G1221-6.

[27] Kim HS, Park DH, Kim JW, Jee MG, Baik SK, Kwon SO, Lee DK. Effectiveness of walking exercise as a bowel preparation for colonoscopy: A randomized controlled trial. *Am J Gastroenterology.* 2005;19:43-50.

[28] Veerabhadrappa SG, Baljoshi VS, Khanapure S, Herur A, Patil S, Ankad RB, Chinagudi S. Effect of yogic bellows on cardiovascular autonomic reactivity. *J Cardiovasc Dis Res.* 2011;2:223-7.

[29] Jerath R, Edry JW, Barnes VA, Jerath V. Physiology of long pranayamic breathing: neural respiratory elements may provide a mechanism that explains how slow deep breathing shifts the autonomic nervous system. *Med Hypotheses.* 2006;67:566-71.

[30] Gaylord SA, Palsson OS, Garland EL, Faurot KR, Coble RS, Mann JD, Frey W, Leniek K, Whitehead WE. Mindfulness training reduces the severity of irritable bowel syndrome in women: results of a randomized controlled trial. *Am J Gastroenterol.* 2011;106:1678-88.

[31] Brower V. Mind-body research moves towards the mainstream. *EMBO Rep.* 2006;7:358-61.

In: Posture: Types, Exercises and Health Effects
Editor: Sarah A. Curran
ISBN: 978-1-63117-252-6

Chapter 12

The Behavior Characteristics and Postural Angles in Teenagers Who Wear High-Heeled Shoes and How Pilates Can Be Used for Postural Control

Patricia Angélica de Oliveira Pezzan, [1,*] ***MSc***
and Daniel Marcondes de Freitas Lopes, [2] ***PT***
[1]MSc. Pontifical Catholic University of Minas Gerais, PUC-MG, Brazil
[2]PT, Strength and Balance Studio, Brazil

Abstract

Body posture in childhood and adolescence undergo changes due to hormonal influences that occur with the onset of puberty and skeletal muscle growth. The pre-pubertal and pubertal phases are crucial to the development of good posture. Among the factors that influence these variables include the use of high-heeled shoes, which are becoming worn more frequently and at an earlier stage within these populations. This chapter aims to explore and discuss the influence of high-heeled shoes on postural angles among adolescents. Using biophotogrammetry, data were collected on 100 adolescents (ages ranging from 13 to 20). Postural angles such as lordosis, horizontal alignment of the pelvis, tibiofemoral angle, angle of the ankle, the Q angle, and angle of the rearfoot were selected. It was perceived that these angles were important in relation to the influence of high-heeled shoes on posture, as well as a reflection on the influence of behavior and clothing characteristics. A secondary aim of this chapter is to present the Pilates method as a focus of treatment and physical practice for patients with spinal disorders. This particular technique describes and emphasizes the need to search for the perfect balance between the body and the mind. It consists of controlling the whole body, whilst the mind

[*] E-mail: patricapezzan@terra.com.br

and breathing enables the focus of the method to be carried out with precision and concentration. In particular, it requires balance and stability from the call center of the body (powerhouse) to perform exercises and control over deep exhalation, with contraction of the rectus abdominis, pelvic floor, and transversus abdominis, as well as keeping the pelvis in a neutral position. This method provides strength and elongation of all muscles of the body, and improves flexibility and coordination by providing relaxation and body awareness. It establishes and enhances appropriate breathing, stimulates circulation and provides correction of posture. For practitioners and clinicians, this method creates a greater awareness of the influence of posture on the body. It also highlights the principles necessary for rehabilitation and maintenance of an ideal posture, functional performance and a healthy pain-free lifestyle.

Keywords: Posture, teenagers, biomechanics, gait, photogrammetry, high-heeled shoes, Pilates

Introduction

Following the adaptation of man to use two limbs to get around and maintain an upright bipedal posture, several physiological changes continued where static posture and spinal muscular action override all gestural actions and dynamic membership. [1-3] The advantage was achieved by the fact that with this assumed posture, the arms became free, and with eyes far away from the ground, there was an increase in the field of vision. In addition, with the release of the arms came the ability to carry and collect food, while the lower limbs began to support the entire weight of body. [4-5]

These changes make a designation of posture, which is a term used to describe not only the alignment of the different body segments, but also describes the orientation of the body through daily life. Correct posture is described as a position, or attitude of the body: an arrangement for the different parts of the body in the formation of a specific attitude, or a specific way someone sustains their body. [6] Therefore, activities undertaken are using less expenditure of energy, which in turn can reduce muscle function. To maintain posture requires a complex neuromuscular system integrated with various proprioceptors in muscles, tendons, joints, and vestibular receptors and visuals. [7]

According to Kendall et al. [8] good posture is the state of muscular and skeletal balance which protects the supporting structures of the body against an injury, or a progressive deformity irrespective of the attitude (i.e., erect, lying, squatting, bent) which these structures are working, or resting.

Magge [9] comments that posture is the composition of the different position of the joints of the body at a given moment and define ideal postural alignment as a straight line (line of gravity), profile view which passes through the earlobe, the body of the cervical vertebrae, the edge of the shoulder joint in the middle of the thoracic spine, the lumbar vertebral body, slightly posterior to the hip joint, slightly above the axis of the knee and just above the lateral malleolus. In this condition, minimum stress is applied to each joint and the least minimal muscle activity is required to maintain the position.

However, many situations of daily life and work may lead to dysfunctions in the entire vertebral column. [10] In addition to these changes, further demands and changes are placed on other body structures. These changes may compromise the development during adolescence and adulthood, as postural patterns acquired during school can be carried out through life. [11]

There are intrinsic and extrinsic factors that can influence the attitude of an individual. These include hereditary, environmental, or physical conditions, socioeconomic status, emotional factors, obesity and physical changes due to development and human growth. [12, 13] Bankoff [14] reported that tonic postural organization is one that is reflected in the attitude of standing posture. This reflects the formation of the body scheme that begins during childhood, through the motor skills acquired at this stage and as a tonic postural imbalance acquired in childhood, adolescence, which can last throughout adulthood.

The incidence of postural deviations in children is significant and is influenced by a variety of factors within the school environment. These include the use of backpacks with weights above the normal weight due to a large quantity of belongings inside, and sitting in a seated position for an extended period of time as well as the use of inadequately sized furniture. [15-20] Recent reviews show an increase of 20% to 51% of lower back pain and other painful occurrences in various musculoskeletal regions. [17, 21-27]

It is observed that the postures of children ranging from the ages of 7 to 12 years suffer changes in the search for new proportions to balance their body and their postural habits. Besides the role and alterations in posture, the feet – the platform to our support are referred to as a complex structure that combines flexibility with stability. [29] As a consequence to alterations and demands placed on posture, the feet may also develop and show changes in terms of structure and function due to repeated bad habits day in and day out.

It is known that the foot is a continuous and evolving structure that does not complete its development until the growing process is complete. [30] It undergoes several changes in its size and shape. At birth, a new born only has an ossified talus and calcaneus, and sometimes the cuboid. [31] With the efforts tasks imposed, the foot passes through structural alterations until it reaches its complete development. The contour of the medial longitudinal arch is the result of adaptability and support which occurs due to foot contact with the ground. This provides balance during static posture and a mechanical advantage for the plantarflexors so that they are able to boost and support the body's weight during the propulsive phase [32] and enable the absorption capacity during impact for various dynamic activities (i.e., walking, running). [29]

Adolescence is described as the transitional stage between childhood and adulthood. It is a process of growth, development, and maturation of the musculoskeletal system that is influenced by internal factors such as hormonal changes and external factors such as heavy tasks (i.e., carrying of rucksacks). This is crucial in the development of good posture during adulthood and changes may influence the development of the locomotive system during puberty, which in turn may lead to postural deviations that becomes irreversible. [33]

At the moment, it is noted that various external factors influence the attitude of adolescents, resulting in an imbalance of the body through which the pathologies arise by postural changes, and among the external factors the influence of high-heeled shoes is noted. The use of which is becoming increasingly common among young adolescents, causing alterations in posture and walking patterns. [15, 34-39] If ignored, changes acquired during childhood may worsen later on during adolescence and adulthood.

Walking patterns may be compromised with increased plantarflexion, associated with the use of high heels. Changes occur in pronation and supination of the foot during walking. [40, 41] With the support area reduced, balance and stability is compromised even during a standing position, whilst during dynamic activities such as walking, an increase in impact forces occur during the initial phase of contact. [41] With this in mind, additional changes and forces to proximal structures of the lower extremities from high-heeled shoes alter the function of the joints of the extremities. [42]

The literature appears to support that the use of high-heeled shoes leads to several problems in the musculoskeletal system. This is mainly due to the compensation mechanisms associated with the lumbar spine, hip, knee, ankle and foot. [43] The presence of lower back pain is reported by many young women, but its' possible cause is shown to be contradictory. However, for some authors [44, 45, 46] this would be a consequence of an increase in curvature of the spine, leading to an overload generated by low back pain as well as the influence of high heels on the musculoskeletal system during walking and other dynamic activities. [47] This however is a broad topic that requires a comprehensive discussion regarding both the causes, and the changes that high-heeled shoes can cause on the body's segments. In addition, such discussion also includes the possibilities of therapeutic techniques for rehabilitation of these changes, as well as approaches to therapies that can be applied with the aim of not only to correct, but also to improve and maintain postural control. In respect to this present chapter, the Pilates method will be discussed as a resource for rehabilitation and training posture.

What Is Posture?

Kendall et al. [8] describes posture as the body's position that involves minimal stretch and stress of its structures with the lowest energy expenditure to achieve maximum efficiency. According to Penha et al. [48, 49] posture is often defined as the relative arrangement of the parts of the body. Penha et al. [49] also commented that good posture is the state of muscular and skeletal balance that protects the structures of the body against injury, or progressive deformity regardless of the attitude in which these structures are working or resting.

Thus, the position reference can be defined by the relationship between the line of gravity body segments. [50] The balance permitted by proper alignment, in turn is defined as the ability to maintain the center of mass of the body that is designed within the boundaries of a supporting structure. [51] Thus, if the body deviates from the balanced condition, restoring forces act on it, seeking to recover balance and stability. [8, 17, 52, 53]

Good posture can therefore be defined as the state of muscular and skeletal balance which protects the supporting structures of the body against injury or progressive deformity irrespective of the condition upright, squatting, lying or bent, in which these structures are working. [8]

The upright posture both in static and dynamic conditions is obtained by the balance between the forces acting on and pulling the body down to the ground, as well as the strength of the posterior muscles of the spine and lower limbs.

Empirically, the reference standard of postural symmetry proposed by Kendall et al. [8] does not occur in the general population, and even people who do not complain of pain in the musculoskeletal system have postural changes. However, there is no standard postural reference that approximates the reality in terms of postural alignment. This creates difficulties in comparing data found in a particular posture assessment of a patient and the evolution of physical therapy with a reference standard that is both reliable and valid.

Poor posture is one where there is a fault in the relationship between various body parts. It produces an increase in stress on the structures supported and results in a less efficient balance of the body and feet. [8] The condition of poor posture can evolve to deformations of physiological curvature of the spine, which affects mainly children and adolescents as well as those in early adulthood. [15] With these changes in the physiological curvature there are alterations in the load axis of the column, whereby the weight distribution is transferred from the disc, which reduces loads to the articular facet. According to Rebelatto et al. [15] this load disturbance can initiate inflammation and pain, favoring the joint degeneration process. These changes may be compromised during adolescence and adulthood, as postural patterns made during the school years acquired, become permanent in adulthood. [11]

In a qualitative study on posture, Penha et al. [49] showed a high incidence of postural changes in children ranging from 7 to 10 years of age. They suggested that these changes may be related to inadequate furniture, sitting habits, and carrying heavy backpacks. These exaggerated loads transported in backpacks are described by Bankoff [14] as an important causative factor of poor posture. Others however acknowledge an additional factor of inappropriate footwear. [11, 14, 15, 49]

All these changes acquired during childhood if left alone and untreated can worsen, and is particularly apparent during adolescence when associated with other factors, one being the use of inappropriate footwear, which has the potential to cause changes in the normal state of the feet. This is due to the fact that during adolescence many postural adaptations occur due to growth. The static postural assessment is important for the understanding of muscle imbalances implicated in many musculoskeletal conditions and is therefore an indispensable tool in clinical practice for the determination of the optimal management strategy. In particular, it specifies which muscles are in a stretched position and which are in the shortened position. [8]

Human Growth and Development

Human growth usually extends from conception until approximately 20 years of age, when growth stops. Structures of the human body reveal physical growth, according to age and vary during different intensities. Both children and adolescents do not grow continuously, but can increase rapidly and then slow down to a steady state until adulthood. Transient acceleration occurs at the time of puberty, which usually occurs between the ages of 11 and 13 years in girls, and the ages of 13 and 15 years in boys. This growth pattern causes a change in body proportion characteristics for each of these periods. [55]

During adolescence, posture undergoes changes because of hormonal influence that occurs with the onset of puberty and skeletal muscle growth. During this period, growth is accompanied by sexual maturation that occurs in women from the ages of 8 to 14 years, and in men from the ages of 9 to 16 years. It is at this stage that the bodily differences between men and women become visible. [9]

These changes in proportion of growth of the body require adjustments to gravity. Pelvic tilt can reduce to approximately twenty five or thirty degrees and the knees are often flexed slightly. For example, primary school children have a period of intense mobility, with the lower age range (from the age of 6), and a longer static, in the range of high age (age 10 for girls and age 11 for boys). This period coincides with the growth spurt. The stabilization of the postural pattern is occurring slowly and adjusting to gravity. [56]

The pre-pubertal and pubertal developments are crucial to the formation and structuring of posture. Due to the necessary adaptations that occur by changing and growth of the body, the body increases load resistance and provides elasticity needed to move. This makes the column more resistant and malleable and therefore, the formation of good posture. In essence, it is during these stages that the vast majority of postural changes arise, resulting in the so called "bad posture," which has as the main causes of poor posture, and bad habits influenced by daily imbalances and muscle dysfunction. [10]

Influence of High-Heeled Shoes on Body Posture

For thousands of years, shoes were used with the primary purpose of protecting the feet during walking and other dynamic activities. [40] However, from the time of Louis XIV, in order to satisfy the desire to become more stylish, women began to wear heeled shoes. [57] Already today it is possible to note that this thinking is increasingly occurring among early adolescents. Thus, the use of high-heeled shoes is present since early time, changing over time, and today can be found in various types of footwear models.

Frey et al. [59] studied the buying habits of shoes among 672 young women (from the ages of 10 to 18 years). They noted that the average age of choosing your own footwear is approximately at the age 8 years, and for 53% of the participants, the style of footwear was the most important factor. High-heeled shoes are popular among women, due to the fact that it is generally considered to produce a more slender and elegant look. [58] The problem is however that many women improperly use these shoes and end up compromising their feet, lower limbs and spinal health.

A study exploring the use of heeled footwear with an average height of a 6.4cm heel noted that the high-heeled shoes favors the displacement of the center of gravity. [35] In this present study, it was noted that the use of high-heeled shoes produced a displacement of the line of gravity, which in turn caused a series of postural adaptations due to this change. The thoracic spine and lumbar spine are effectively positioned more towards the line of gravity, whilst the knee is more flexed, with an increase in ankle plantarflexion.

This observation is explained by Opila et al. [35] who compared postural alignment in a barefoot condition and high heeled shoes. They noted that during the time of use for the test, there was momentary compensation for maintaining balance, and as such, the lumbar lordosis present flattened, and this was associated with a posterior pelvic tilt. However, this finding is contrary to those found in personal studies and according to the author, there is no compensation. This is because the abdominal muscles and posterior column become fatigued and the tone is reduced due to an extended period of time in this posture. It is possible that with repeated use of this type of shoe the column will become hyperlordotic. This same author also studied the displacement of the center line of gravity, comparing the barefoot condition with high-heeled shoes, and noted that this footwear line with the center of gravity is closer to the lateral malleolus at about 6 mm. Thus, it was possible to analyze and quantify changes in compensation that the various body segments acquired.

The use of high-heeled shoes reduces the support of the feet, the platform of the body, which can proportionally increase the purchase and contact with the ground for the metatarsal heads and toes. It was also noted by Opila et al. [35] and others [36, 37] that the use of high-heeled shoes position the feet in a more plantarflexed position. In this position, the bodyweight is transferred to the forefoot and triceps surae which is more shortened and results in a greater ability to develop less contractile force.

The increase in forefoot overload is responsible for the frequency of reported pain in the feet. [44, 45, 46] This can lead to the appearance of calluses under the metatarsal heads, especially the second through to fourth. Another cause for the appearance of calluses is the collapse of one or more plantar arches. [35, 37, 38] These statements bring the following questions:

1. If the incentive for the use of a fad (i.e., high heels) is presented at an earlier age, and is not questioned for the detrimental effects they can have on postural control, what is the future of our teenagers?
2. Is it more appropriate to have adults and elderly women with a good quality of life that provides physically and functional independency, with less pain and degeneration? This creates harmony between body and mind, which is supports the great pursuit of health in modern day living.

These reflections are important and as clinicians we should not only reflect, but intervene to enable this reality to change.

Users and Non-Users of High-Heeled Footwear

In order to discuss the behavioral characteristics and postural angles of users of high-heeled shoes, the author presents findings of previous research. One hundred adolescent females, aged 13 to 20 years, who attended private schools and universities in Brazil, were invited to take part. They were divided into two groups: 50 adolescents in the non-users of high-heeled shoes (NUG) and 50 adolescents in the group of users of high-heeled shoes (UG). Below are the survey data and discussion of results. [61]

1. Habitual Trends with High-Heeled Shoes

The findings showed that the most used shoes for adolescent wearers were the heel footwear soled drive type Anabella. It was also observed that high-heeled shoes that are thick, thin and platform are also used. However, more adolescents (non-users) are using use tennis type shoes / sneakers more often. It was noted that that they all eventually use high-heeled shoes. The average height of high-heeled shoes used by all adolescents was 9.8 cm.

The group of non-users had an average frequency of use of 1.74 times per week, with an average weekly use of 15 hours. This was smaller than the frequency found in the group of high-heel users who had an average of 4 times week usage and an average of 30 hours on a weekly basis.

From this data it was therefore observed that the majority of adolescents used high-heeled shoes at some other time during the week, even the teenagers classified as non-users. This data is somewhat concerning since the use of these types of shoes has been shown early among this population, young adolescents are now making use of high-heeled shoes, and the hours spent using these shoes is increasing. This contributes undoubtedly to the increase in postural problems and biomechanical gait changes becoming more evident.

2. Analysis of Lumbar Spinal Postural Angles, Pelvis and Lower Limbs

This analysis involved the use of postural photogrammetry in two conditions:

1. With no shoes on;
2. With shoes on.

With shoes on, the study was conducted wearing the footwear known as Anabella and was worn for a period of one hour to enable postural adaptations could take place. The postural angles measured and their interpretation are presented in table 1. Comparison between groups for each variable studied in both conditions (without and with high heels; tables 2, 3 and 4).

Table 1. Interpretation of postural angles

Postural Variables	Interpretation of the values	Interpretation of the values
Lumbar lordosis angle	The larger the angle the greater the rectification	The smaller the angle the greater the hyperlordosis
Horizontal alignment of the pelvis	The larger the angle the greater the anteversion	The smaller the angle the greater the reversion
Angle of the knee (posterior angle)	Greater than 180° - hyperextension knees	Less than 180° - half knee flexion
Angle tibial tarsal (anterior angle)	Greater than 90° (ankle plantarflexion)	Less than 90° (ankle dorsiflexion)
Front angle of the lower limb right and left	Values greater than 175° - Knee varus	Values less than 170° - knee valgus
Q angle (valgus angle patellar) right and left	Greater than 18° - patella valgus	Less than 15° - patella varus
Rearfoot angle right and left	Values greater than 0° - rearfoot valgus	Values less than 0° - rearfoot varus

Table 2. Mean, standard deviation and p values comparing the conditions with and without footwear Anabela group non-users (NUG) and group users (UG) *Significant difference

Variable	Without and with the heel	Mean (NUG) (n = 50) (degrees)	Mean (UG) (n = 50) (degrees)	P (NUG)	P (UG)
Angle of lumbar lordosis	without	62 ± 4.30	40 ± 5.30	<0.001*	<0.001*
	with	79 ± 3.20	38 ± 4.10		
Horizontal alignment of the pelvis	without	-9.57 ± 6.78	-14.57 ± 4.48	<0.001*	<0.001*
	with	2.01 ± 7.37	-16.62 ± 6.85		
Knee angle	without	181.13 ± 4.77	180.8 ± 7.56	0.764	0.773
	with	182.75 ± 4.33	182.90 ± 6.16		
Angle tibial tarsal	without	94.72 ± 2.99	84.61 ± 3.20	<0.001*	<0.001*
	with	112.08 ± 5.75	112.25 ± 4.92		
Front angle of the lower limb right	without	168.37 ± 4.42	165.11 ± 2.62	<0.001*	<0.001*
	with	163.27 ± 2.48	163.45 ± 2.75		
Front angle of the lower limb left	without	168.21 ± 4.10	165.41 ± 3.20	<0.001*	0.058
	with	163.83 ± 3.54	163 ± 3.15		
Q angle right	without	12.50 ± 2.64	20.50 ± 3.54	<0.001*	<0.001*
	with	18.51 ± 2.42	23.68 ± 3.26		
Q angle left	without	12,23 ± 2,27	20.70 ± 3.13	<0.001*	<0.001*
	with	17.85 ± 2.32	22.00 ± 2.80		
Rearfoot angle right	without	11.41 ± 4.39	3.35 ± 3.10	<0.001*	<0.001*
	with	-2.49 ± 2.80	-4,76 ± 3,43		
Rearfoot angle left	without	10.84 ± 3.68	3.34 ± 2.97	<0.001*	<0.001*
	with	-2.45 ± 3.07	-4.66 ± 3.51		

Reprinted with permission from Pezzan PAO, João SMA, Ribeiro AP, Manifio EF. Postural assessment of lumbar lordosis and pelvic alignment angles in adolescent users and nonusers of high-heeled shoes. *J Manipul Physiol Therap.* 2011;34:614-21.

This study showed that high-heeled shoes (Anabella) is a causative factor in differences in postural parameters between both the conditions of use as between groups. It was observed that the angle of lumbar lordosis behaves differently between groups and between conditions, although both presented in the barefoot condition a posture lumbar lordosis with a mean of 62° ± 4.30 for the group of adolescents non-users and 40° ± 5.30 for the group of users, the adolescent wearers have a steeper angle of concavity than non-users.

Table 3. Comparison of variables between the non-users (NGU) and group users (UG) for the condition without footwear Anabella. *Significant difference

Barefoot		Mean (degrees)	n	p-value
Angle of lumbar lordosis	NGU	62 ± 4.30	50	0.001*
	UG	40 ± 5.30	50	
Horizontal alignment of the pelvis	NGU	-9.57 ± 6.78	50	0.001*
	UG	-14.57 ± 4.48	50	
Knee angle	NGU	181.13 ± 4.77	50	0.103
	UG	180.8 ± 7,56	50	
Angle tibial tarsal	NGU	94.72 ± 2.99	50	0.041*
	UG	84.61 ± 3.20	50	
Front angle of the lower limb right	NGU	168.37 ± 4.42	50	0.006*
	UG	165.11 ± 2.62	50	
Front angle of the lower limb left	NGU	168.21 ± 4.10	50	0.006*
	UG	165.41 ± 3.20	50	
Q angle right	NGU	12.50 ± 2.64	50	<0.001*
	UG	20.50 ± 3.54	50	
Q angle left	NGU	12.23 ± 2.27	50	<0.001*
	UG	20.70 ± 3.13	50	
Rearfoot angle right	NGU	11.41 ± 4.39	50	<0.001*
	UG	3.35 ± 3.10	50	
Rearfoot angle left	NGU	10.84 ± 3.68	50	<0.001*
	UG	3.34 ± 2.97	50	

Reprinted with permission from Pezzan PAO, João SMA, Ribeiro AP, Manifio EF. Postural assessment of lumbar lordosis and pelvic alignment angles in adolescent users and nonusers of high-heeled shoes. *J Manipul Physiol Therap*. 2011;34:614-21.

However, by placing the heel, non-users inverted posture of the lumbar lordosis acquiring a rectification with an average of 79° ± 3.20, since the adolescent wearers only had an increased lumbar hyperlordosis of an average of 38° ± 4.10.

There is much controversy in the literature regarding the stance taken by the lumbar spine when influenced by high heels, as well as the position of the pelvis. Studies by Opila et al., [35] Bendix et al., [62] and Franklin et al. [63] reported the occurrence of a decrease in lumbar lordosis with increasing heel height and assumed the posture of lumbar rectification corroborate the findings presented within this chapter with regards to the group of non-users that showed rectification of the lumbar lordosis. The authors believed that the correction occurred in these adolescents seeking compensation in postural balance. Since the heel shifts the center of gravity above and causes postural imbalance, in order to resume it, it's possible that adolescents use a compensatory strategy assuming a posture of lumbar rectification. However, the group of users showing different results found by these authors, suggest an increased lumbar lordosis. The group already makes use of this footwear and its' chronic use has its' influence over time has led to an adaptation of the lumbar spine in hyperlordosis.

Table 4. Comparison of variables between non-users (NGU) and group users (UG) for the condition with footwear Anabella. *Significant difference

High-heeled		Mean (degrees)	n	p-value
Angle of lumbar lordosis	NGU	79 ± 3.20	50	<0.001*
	UG	38 ± 4.10	50	
Horizontal alignment of the pelvis	NGU	2.01 ± 7.37	50	<0.001*
	UG	-16.62 ± 6.85	50	
Knee angle	NGU	182.75 ± 4.33	50	0.391
	UG	182.90 ± 6.16	50	
Angle tibial tarsal	NGU	112.08 ± 5.75	50	0.874
	UG	112.25 ± 4.92	50	
Front angle of the lower limb right	NGU	163.27 ± 2.48	50	0.119
	UG	163.45 ± 2.75	50	
Front angle of the lower limb left	NGU	163.83 ± 3.54	50	0.120
	UG	163 ± 3.15	50	
Q angle right	NGU	18.51 ± 2.42	50	0.001*
	UG	23.68 ± 3.26	50	
Q angle left	NGU	17.85 ± 2.32	50	<0.001*
	UG	22.00 ± 2.80	50	
Rearfoot angle right	NGU	-2.49 ± 2.80	50	0.001*
	UG	-4.76 ± 3.43	50	
Rearfoot angle left	NGU	-2.45 ± 3.07	50	0.001*
	UG	-4.66 ± 3.51	50	

Reprinted with permission from Pezzan PAO, João SMA, Ribeiro AP, Manifio EF. Postural assessment of lumbar lordosis and pelvic alignment angles in adolescent users and nonusers of high-heeled shoes. *J Manipul Physiol Therap.* 2011;34:614-621.

This finding is supported by studies from Nasser, [64] Kulthanan et al. [65] and Snow et al. [66] in describing that high-heeled shoes increases lumbar lordosis.

Lumbar spine posture is directly related to pelvic postures, and therefore, the literature describes the changes in lumbar lordosis in conjunction with changes in the pelvic segments. Thus, it was observed that both groups had pelvic anteversions accompanied by lumbar hyperlordosis in the barefoot condition. However, the group of users demonstrated more accentuated pelvic anteversion angles. While maintaining the use of high- heeled shoes, the group of non-users adopted compensatory patterns with a reversal condition of the pelvic posture, which was placed in retroversion. It is believed that this occurred because high-heeled shoes moved the center of gravity anteriorly, and therefore, there was a need to regain balance. This resulted in the tilting of their pelvis posteriorly, just as with lumbar rectification. These data corroborated those reported by Manfio et al. [67] and Franklin et al. [63] These authors concluded that high- heeled shoes would reduce pelvic anteversion, and the barefoot condition would increase this angle. [67] It is important to note that they investigated the acute effects of these types of shoes, as was the case with the group of non-users in the present study. However, the group of users had different results, in which there were observed increases in the anteversion angles with the use of high-heeled shoes. This may have happened because these participants were already users of this type of footwear; they kept an adaptive posture of anteversion by having their center of gravity already displaced because of the prolonged use of high heels. These findings were also found by Nasser, [64] Kulthanan et al., [65] and Snow et al. [66]

The angle of knee flexion and extension was not different between conditions or between groups, a result also obtained by Aguiar and Freitas [68] and Iunes et al. [69] As expected, the angle for dorsiflexion and plantarflexion is different between conditions. For with and without (barefoot condition) footwear, it was noted for ankle plantarflexion that the barefoot condition had an average of 94.72° ± 2.99 in the non-users group and 84.61° ± 3.20 for the users. This increased the average to 112.08° ± 5.75 and 112.25° ± 4.92 respectively. Among the groups, it was found that the value was significantly different only in the barefoot condition, where the group of users and non-users have a value less than 90°, but the users have a lower value compared to non-users. Possibly, depending on the time of use of high-heeled shoes and the constant maintenance of the displacement of the center of gravity above, the users even when they are not with high-heeled, continue with the center of gravity shifted earlier, reducing the tibial tarsal angle.

The frontal plane angle of the lower limb presented with a valgus provided without high-heel shoes or both groups and after staying with the footwear Anabella. This value significantly increased to valgus, with the exception of the left knee with the group of users showing after the introduction of the high-heeled shoes, the angle was marginally different, with it displaying a tendency to be higher in this condition.

Both adolescent non-users and users presented with a valgus at the knee, and this was more pronounced in users. The variation however was greater in non-users, approaching the values of valgus groups being provided with heels. This justifies the fact of non-users and users alike are using the footwear. Moreover, these findings indicate that the frequent use of this footwear increases the valgus knee, which is already described as a characteristic posture of knees in women, as well as a greater Q angle (valgus angle of the patella). This condition may cause biomechanical changes overtime which in turn generate a progressive loss of cartilage and pain due to overload of the lateral compartment of the tibiofemoral joint. [70]

From a dynamic point of view, this increased angle can increase the strength of the lateral force vector (valgus). [71, 72, 73] This enhances the lateral movement of the patella, resulting in pain, which can be aggravated when people flex their knee in physical activities such as going up and down a flight of stairs, sitting for a prolonged time, squatting, kneeling, and also in physical exercises. Furthermore, there may be a sensation of cracking and popping, "buckling" of the knee.

The adolescent non-users in this study showed a Q angle of 12.50° ± 2.64 (right) and 12.23° ± 2.27 (left) in the condition without shoes, and is considered to be close to the value considered normal (15° to 18°). [70] For users, the value was higher than normal in the same condition with 20.50° ± 3.54 (right) and 20.70° ± 3.13 (left), showing that the adolescents in this group, even when they are not under the acute effects of high heels, already have a valgus patella. After using the footwear, both groups showed a significantly increased angle however, again as with the valgus knee, the increase was greater in the group with non-users 18.51° ± 2.42 (right) and 17.8° ± 2 32 (left), while users in the group to the increase was 23.68° ± 3.26 (right) and 22.00° ± 2.80 (left). It is acknowledged that the angle increased, but with less variation than for non-users, and may be due to the valgus position these users already have from wearing high-heeled shoes. Even with a greater variation of this angle in the non-users, the groups remain statistically different when wearing the Anabella footwear.

The increase of the angles mentioned above (frontal angle of the lower limb and Q angle) by the influence of high-heeled shoes and their use may trigger the occurrence of osteoarthritis [42] and meniscal and ligamentous instability. This may result in physical changes of the knee, and joint over a period of time.

By analyzing the position of the feet, it was observed that both the users of high-heeled shoes and non-users displayed a rearfoot valgus angle, in the condition without footwear with 11.41° ± 4.39 (right) and 10.84° ± 3.68 (left) for non-users and 3.35° ± 3.10 (right) and 3.34° ± 2.97 (left) for users. This appears to be a condition which tends to generate a collapse of the arch (i.e., flat feet) which in this case was more present in non-users by demonstrating a greater rearfoot valgus. It was observed that the users group had a significant value of valgus that was lower compared to non-users, significantly. This finding suggests that these users had a compensatory strategy to adapt postural positioning, and with the time of use, the feet were closer to 0° for the rearfoot angle, which in turn, resulted in a value of varus greater for this group for the Anabella (with heel) condition to - 4.76 ° ± 3.43 (right) and - 4.66 ° ± 3.51 (left). This may explain the fact that although both groups after staying with footwear Anabella, have significantly modified the rearfoot angle, and it assumed a varus position whereby users showed a varus greater than non-users. Again this favors change in the plantar arch which now will increase (i.e., supinated).

The high-heel shoes maintain a posture of the ankle and foot in a shortened position of the extensors and inverters. This may explain the posture assumed by the varus rearfoot to stay within the high-heels, as well as the attempt to achieve better stability and balance. Moreover, according to Gastwirth et al., [41] the wearing of high-heel shoes in females limits pronation of the subtalar joint. According to these authors, this may be related to the emergence of problems in the knee, hip and lumbar spine, and also modification of the plantar arch. These findings lend support to the descriptions of Snow and Williams, [36] who after studying the angle of pronation in women's shoes with different heel heights, indicated that the occurrence of increased rearfoot pronation with wearing low-heeled shoes compared with medium and high-heeled shoes. According to these authors, supination maintained for a long time causes stretching of the ligaments of the lateral compartment of the ankle and foot, [36] which could result in increasing the risk of instability. [74]

This evidence suggests that the constant use of high-heeled shoes, especially when started in the early growth phase, may lead to inappropriate postural alignment of the feet, which eventually culminates in a lower efficiency in the mechanics of movement.

3. Summary

The prolonged use of high-heeled shoes or chronic use of this type of footwear worn since adolescence can disrupt good posture by creating the following:

- An increased lumbar lordosis and a pelvis which is positioned in anteversion;
- By keeping the center of gravity above the high-heeled shoes because, over time use, postural adaptations that make the users even when they are not with footwear, maintain the forward center of gravity and remain with tibial tarsal angle less than 90°;
- The position of the knee in the frontal plane and the Q angle becomes valgus (increases) with use of high heels;
- Chronic use of high heels causes the rearfoot to assume a varus position.

This research showed that high-heeled shoes, especially Anabella negatively influenced the posture of the lumbar spine, pelvis and lower limbs. This affects postural alignment which can lead to a range of various compensatory mechanisms that can occur during the growth phase. Future problems can include pain, muscle and joint dysfunction, and can result in functional limitations as well as the loss of body awareness.

Whilst the target population of this study was girls who were in the growth phase, the detections of these postural data can support preventive and therapeutic procedures especially in this population. This chapter will now explore how this approach could correct and prevent problems that may arise, through the use of the Pilates method.

Pilates and Postural Control

The Pilates method, initially known as *"The Art of Contrology"* was created by Joseph Hubertus Pilates. [75] This method is based on strength exercises and mobility through specific exercises developed, and it brings many benefits to the body such as:

- Strengthening and stretching the entire body musculature;
- Improving flexibility and coordination;
- Providing relaxation;
- Bringing greater body awareness;
- Maintaining constant breathing;
- Stimulating circulation;
- Optimizes the performance of athletes;
- Provide correction of posture. [76]

It is a method that develops a greater awareness of the body predominantly through the following principles: concentration, control and center force (powerhouse) which are defined by the abdomen, lower back and buttocks. These should remain activated during the whole time. This provides further stabilization of the back, allowing the body during all movements, axial elongation, prevention of lower back pain, fluid movement, breathing and accuracy (characterized by a work of breathing and control of the thoraco-abdominal).

A difference of this method relates to the importance that the recovery or maintenance of postural balance is due to the fact that the Pilates exercises advocates the following: movements performed particularly during expiration, with the contraction of the transversus abdominis, the main stabilizer of the trunk especially at lumbar spine and lower ribs, preventing injuries and pain in the same, the neutral position of the pelvis, allowing optimal alignment and consequently lordosis lumbar; knee extension position, which results in a constant stimulation of balance. Therefore, it is acknowledged that this is an interesting method both in rehabilitation resulting from physical activities.

Netto Colodete [77] stated an important advantage for this method noting the fact of using the springs that make up the equipment of Pilates. They have various lengths and diameters, offering diverse changing patterns of resistance and assistance in the progression of their exercises. Several studies have shown that the nature of the spring force is not constant and changes with the variation of the length (the strength of the force varies linearly with the distance the spring is stretched or compressed relative to its initial length), and the greater the change in its' length, the greater the strength. Furthermore, a peak of strength is required so that the spring resistance is surpassed (law of inertia), so that there is a gradual increase of strength.

The linear power gain is given by the fact that after initiating movement with a peak force which occurs according to the deformation of springs, the resistance increases gradually. It also increases the need for generating force to overcome this resistance gradually, with the prime mover muscle initially active. With the progression of resistance, the need to increase the strength of the secondary motor is then activated. Given this, it is understood that the practice of Pilates improves strength in linear physical activities and sports. Therefore, athletes who require greater force to linear performance can benefit from this practice. Another distinguishing feature of the Pilates method is strengthening global. That way you can work the body through the kinetic chain and muscle together, showing that to have sound, mind, and body with muscular endurance and flexibility through mindfulness of body movements are necessary to enable all complex biomechanical and kinesiological patterns.

When one considers preventing or correcting postural dysfunctions it is initially thought that without an overall body and mainly sensory motor, we can hardly change the unbalanced body awareness in these conditions. This is perhaps the most prominent reason why this Pilates method can bring many advantages to posture and awareness. Important kinesiological and biomechanical advantages as related to Pilates are:

- Decreased joint impact, because the maintenance of the principle of axial growth structures is removed and that the friction decreases protecting joints against wear;
- Provides more proprioceptive motor engram;
- Progressive resistance;
- Enables first prime movers;
- Gain dynamic stretching;
- Constant and balanced work the abdominals, stabilizers and erector spine;
- Increases linear force.

The relationship between abdominal muscle work and the erector spine, and the work of the whole body maintains an optimal biomechanical by balancing between force and muscle stretching, flexibility and body awareness. There is convincing evidence that the realization of an exercise program with emphasis on strengthening the trunk extensor musculature restores the function of the lumbar spine and can prevent the onset of back pain. In a survey conducted by Kolyniak et al., [78] the Pilates exercises with intermediate-advanced was efficient to promote increased peak torque, total work, average power and total work of muscles related to extension trunk. These results indicated that this method of training can be used as a strategy for strengthening these muscles, reducing the imbalance between the muscles involved in the extension and trunk flexion. This would appear to make the method a system of treatment and physical training biomechanically less harmful to the body segments, especially the spine.

Kloubec [79] showed that after a training period of 12 weeks, participants showed a statistically significant increase in their height. This indicated that some changes may have occurred in the structural alignment of the spine, while the postural analysis used was not sophisticated enough to detect discrete changes. Nevertheless, this is an important result that requires more attention and research. What these findings do show however, is that you can determine postural orthostatic alignment of the body and then measure height and once Pilates is acted on possible misalignment of the spine can infer that it may be beneficial interventions postural correction. It is worth noting that although this research has not proposed a series of specific exercises for posture correction, it did find positive results for rebalancing the same. Moreover, practice has shown that Pilates was able to produce a statistically significant increase in abdominal resistance, flexibility, and resistance in hamstring muscles in the upper body.

The effectiveness of Pilates to improve postural aspects can be also supported by the results found by Emery et al. [80] They suggested that post workout Pilates could improve strategies for stabilization of the thoracic spine and thus better control of bilateral scapulae trunk and the rebalancing of the abdominal muscles and scapular stabilizers. This in turn results in a significant reduction in thoracic kyphosis. With these observations, one can expect changes in the cervix and activities of the erector muscle of spine. Thus, Pilates can facilitate postural symmetry and prevent the risk of injury during body movements supporting the work of Muscolino and Cipriani [81] who state that this is a method that focuses primarily on the center of strength and postural control. The discussions and conclusions of these authors lend support to suggest that Pilates as a form of correction and postural control in patients with these disorders including users of high-heel shoes (as shown earlier in this chapter) have important changes in different body segments.

The research cited also supports clinical evidence regarding a better postural alignment observed in students of Pilates. It is noted that those who are referred for treatment and postural correction, and those who join to practice Pilates as a method of physical training can correct postural problems. In particular, Pilates has a series of principles which exercises through the control of force stabilizing muscles of the lumbar spine, deep breathing with activation of abdominal axial growth. This favors decompression of the joints, especially the spine, pelvis in a neutral position and constant work with sensorimotor body awareness and balance. These are variables that should be included in any postural work and physical activities performed using these variables could provide better harmony with full body balance between the forces responsible for the biomechanics of posture and movement throughout the body. This enables the integration of balance between the body and mind.

Conclusion

It is concluded that through the body, work under vision therapy or the physical training of humans can enhance their movement with less energy expenditure and reduce pain, symptoms and functional limitations. This in turn can lead to a better quality of life and physical independence. This chapter has explored the need to recognize the effects of high-heeled footwear on posture in teenagers and minimize postural and functional compensations. The use of Pilates may also serve as a useful method of assisting postural stability and strength which could minimize the risk of occurrence of musculoskeletal pathologies in the future.

References

[1] Souza JL, Lemos TV, Luiz MMM. Método Mackenzie vs Williams: uma reflexão. *Rev Fisioter Bras.* 2003; 4: 67-8.

[2] Lewin R. Evolução humana. São Paulo - SP: Atheneu; 1999,

[3] Toni PM, Salvo CG, Marins MC, Weber LND. Etiologia humana: o exemplo de apego. *Psico-USF.* 2004; 9: 99-104.

[4] Borges FR, Almeida SJA. Locomoção humana: diretrizes terapêuticas com base nos conhecimentos evolutivos. *Arq Ciênc Saúde.* 2004;11:72-5.

[5] Shumway-Cook, A. Woollacott MH. Motor-control: Theory and practical applications. 3rd ed. Philadelphia: Other Editions; 2000.

[6] Kisner C , Colby LA. Exercícios terapêuticos. São Paulo (SP): Editora Manole; 1987.

[7] Ferrario VF, Sforza C, Tartaglia G, Barbini E, Michielon G. New television technique for natural head and body posture analysis. J Craniomandib Pract. 1995;13:247-55.

[8] Kendall FP, McCreary EK, Provance PG. Múculos: Provas e Funções. 5ª ed. São Paulo (SP): Editora Manole;1995.

[9] Magee DJ. Orthopedic Physical assessment. 3ª ed. Philadelphia: Editora Saunders Company; 1997.

[10] Barbosa L, Braga ES, Frederico BR, Madeira JS. Prevalência de lombalgia em acadêmicos de fisioterapia no ambulatório de um hospital universitário. *Rev Fisioter Bras.* 2002;3:372-3.

[11] Neto FR. Avaliação postural em escolares de 1ª à 4ª série do 1° grau. *Rev Bras Ciênc Mov.* 1991;5:7-11.

[12] Asher C. Variações de postura na Criança. São Paulo (SP): Editora Manole; 1976.

[13] Rodrigues RL, Barbanti VJ. Atividade esportiva e a criança. Principais lesões do aparelho locomotor. In: Conceição JAN. Saúde escolar: a criança, a vida e a escola. São Paulo (SP): Editora Sarvier; 1994.

[14] Briguetti V, Bankoff ADP. Levantamento da incidência de cifose postural e ombros caídos em alunos de 1ª à 4ª série escolar. *Rev Bras Ciênc Esporte.* 1986;7:93-7.

[15] Rebelatto JR, Caldas MAJ, Vitta A. Influência do transporte do material escolar sobre a ocorrência de desvios posturais em estudantes. *Rev Bras Ortop.* 1991;26: 403-10.

[16] Casarotto RA, Liberti EA. Dados antropométricos de pré-escolares da cidade de São Paulo. *Rev Fisioter Univ São Paulo.* 1994;1:1.

[17] Watson KD, Papageorgiou AC, Jones GT, Taylor S, Symmons DP, Slman AJ, Macforlane GJ. Low back in schoolchildren: occurrence and characteristics. *Pain.* 2002;97:87-92.

[18] Lafond D, Descarreaux M, Normand MC, Harrison DE (2007). Postural development in school children: a cross-sectional study. Chiropr Osteopat [On -line série]. 15:1. Available fron: URL: http://www.chroandosteo.com/content/ 20/10/2007.

[19] Murphy S, Buckle P, Stubs D. Classroom posture and self-reported back and neck pain in schoolchildren. *Appl Ergon.* 2004;35:113-20.

[20] Cardon G, De Clercq D, De bourdeaudhuji I, Breithecker D. Sitting habits in elementary schooldchildren: a traditional versus a "Moving school". *Patient Educ Couns.* 2004;54:133-42.

[21] Viry P, Creveuil C, Marcelli C. Nonspecific back pain in children: a search for associated factors in 14-year-old schoolchildren. *Rev Rhum Engl Ed.* 1999; 66: 381-8.

[22] Balague F, Troussier B, Salminen JJ. Non-specific low back pain in children and adolescents: risk factors. *Eur Spine J.* 1999; 8: 429-38.

[23] Olsen TL, Anderson RL, Dearwater SR, Krisk AM, Cauley JA, Aoron Dj, La Porte RE. The epidemiology of low back pain in an adolescent population. *Am J Public Healthy.* 1992; 82: 606-8.

[24] Taimela S, Kujala VM, Saminen JJ, Viljanen T. The prevalence of low back pain among children and adolescents: a nation-wide, cohort-based questionnaire survey in Finland. *Spine.* 1997; 22: 1132-6.

[25] Cardon G, Balague F. Low back pain prevention's effects in schoolchildren: what is the evidence? *Eur Spine J.* 2004; 13: 663-79.

[26] Parcells C, Stommel M, Hubbard RP. Mismatch of classroom furniture and student body dimensions: empirical findings and health implications. *J Adolesc Health.* 1999; 24: 265-73.

[27] Marschale M, Harrington AC, Steele JR. Effect of Work station design on sitting posture in young children. *Ergonomics.* 1995; 38: 1932-40.

[28] Fernandes M, Rodrigues LF, Barros JW, Shimano AC, Moreira FBR, Gonçalves FF, Amorim GS, Otoni NT, Hackenberg L, Hierholzer E, Pötzl W, Götze C, Liljenqvist U. Rasterstereographic back shape analysis in idiopathic scoliosis after anterior correction and fusion. *Clin Biomech.* 2002;18: 1-8.

[29] Norkin CC, Levangie PK. Articulações: Estrutura e Função – Uma Abordagem Prática e Abrangente. In: Marcha. 2nd ed. Rio de Janeiro (RJ). Editora Revinter; 2001. p. 399-404.

[30] López N, Albuquerque F, Santos M, Sánches M, Dominguez R. Evaluation and analysis of the footprint of Young individuals: A comparative study between football players and non-players. *Eur J Anat.* 2005;9:135-42.

[31] Viladot A. "Malformaciones del dedo gordo" in Patologia del antepié. Barcelona. Ediciones Toray; 1981.

[32] Morton DJ. Evolution of the longitudinal arch of the human foot. *J Bone Joint Surg.* 1924; 6:56-90.

[33] Setian N. Endocrinologia Pediátrica: Aspectos Físicos e Metabólicos do Recém Nascido ao Adolescente. In: Saito MI. Padrões do Desenvolvimento Pubertário e suas Variações. 2° ed. Editora Savier; 2002.

[34] Grimmer KA, Williams MT, Gill TK, M App Sc. The associations between adolescent head-on-neck posture, backpack weight, and anthropometric features. *Spine.* 1999; 24:2262-72.

[35] Opila KA, Stephen SW, Stanley Schiowitz, Chen J. Postural alignment in barefoot and high-heeled stance. *Spine.* 1988;13:542-7.

[36] Snow RE, Williams KR. High heeled shoes: Their effect on center of mass position, posture, three-dimensional kinematics, rearfoot motion and ground reaction forces. *Arch Phys Med Rehabil.* 1994;75:568-76.

[37] Opila-Correia KA. Kinematics of high-heeled gait. *Arch Phys Med Rehabil.* 1990;71: 304-9.

[38] Lateur BJ, Giaconi RM, Questad BA, Ko M, Lehmann JF. Footwear and posture: Compensatory strategies for heel height. Am. J. Phys. Med. Rehabil. 1991;70:246-54.

[39] Júnior ASA, Freitas TM. Biomecânica da marcha e da postura com calçado de salto alto. *Rev Fisioter Bras.* 2004; 5:183-87.

[40] Frey CMD. Foot health and shoewear. *Clin Orthop Relat Res.* 2000; 372: 32-44.

[41] Gastwirth BW, O`Brien TD, Nelson RM, Manger DC, Kinding SA. An electrodynographic study of foot function in shoes of varying heel heights. *J Am Podiatr Med Assoc.* 1991; 81: 463-72.

[42] Kerrigan DC, Todd MK, Riley PO. Knee osteoarthritis and high-heeled shoes. *Lancet.* 1998; 351: 1399- 401.

[43] Buehler VL. The Effect of various heights of heels upon erect body posture and an investigation of possible reasons [thesis]. Eugene, OR: University or Oregon; 1932.

[44] Jang N, Kim Y. Human physiology, 3ª Ed. Seoul, Korea: Sumunsa; 1998.

[45] Yoe H. A study on the compatibility of women's′ shoes [thesis]. Pusan University Pusan (Korea); 1994.

[46] De Luca CJ. Low Back Pain: A major problem with low priority. J Rehabil Res Develop. 1997;34:7-8.

[47] Pegoretti C. Adaptações das curvas da coluna vertebral na marcha em função da altura dos saltos dos calçados [tese]. São Paulo (SP). Universidade São Paulo; 2004.

[48] Penha PJ, Baldini M, João SMA. Spinal postural alignment variance according to sex and age in 7-and 8-year-old children. *J Manipulative Physiological Ther.* 2009;32:154-9.

[49] Penha PJ, João SMA, Casarotto RA, Amino CJ, Penteado DC. Postural assessment of girls between 7 and 10 years of age. *Revista Clinics.* 2005; 60: 9-16.

[50] Gangnet N, Pomero V, Dumas R, Skalli W, Vital J-M. Variability of the spine and pelvis location with respect to the gravity line: a three-dimencional stereoradiographic study using a force platiform. *Surg Radiol Anat.* 2003;25:424-33.

[51] Shumway-Cook A, Woollacott MH. Motor-control: theory end practical applications. Maryland (USA): Lippincott Williams and Wilkins; 2001.

[52] Christie HJ, Kumar S, Warren SA. Postural aberrations in low back pan. *Arch Phys Med Rehabil.* 1995;76:218-24.

[53] Zatsiorsky VM, Duarte M. Instant equilibrium point and its migration in standing tasks: rambling an trembling components of the stabilogram. *Motor Control.* 1999;3: 28-38.

[54] Tanaka C, Farah EA. Anatomia funcional das cadeias musculares. São Paulo (SP): Ícone; 1997.

[55] Rodriguez MD, Sacco ICN, Amadio AC. Estudo biomecânico do índice do arco longitudinal plantar em crianças de diferentes grupos experimentais. Anais do VII Congresso Brasileiro de Biomecânica; 1997; Campinas; Brasil.
[56] Knoplich J. Enfermidades da Coluna Vertebral – Uma visão clínica e fisioterápica. 3ª Ed. São Paulo (SP): Robe editorial; 2003.
[57] Linder M, Saltzman CL. A history of medical scientists on high heels. *Int J Health Serv.* 1998;28:201-25.
[58] Réssio C. O Preço da Elegância [serial online] 2000. Available from: URL: http://www.sabido.com.br/artigo.asp. Accessed 14th November 2013.
[59] Frey C, Thompson F, Smith J, Sanders M, Horstman H. American Orthopaedic Foot and Ankle Society Women's Shoe Survey. *Foot Ankle.* 1993;14:78-81.
[60] Linder M, Saltzman CL. A history of medical scientists on high heels. *Int J Health Serv.* 1998; 28: 201-25.
[61] Pezzan PAO, João SMA. Avaliação postural da coluna lombar, dos membros inferiores, e análise da força reação do solo em adolescentes usuárias de calçados de salto alto [dissertação]. São Paulo (SP): Universidade São Paulo - USP; 2009.
[62] Bendix T, Sorenson SS, Klausen K. Lumbar curve, trunk muscles and line of gravity with different heel heights. *Spine.* 1984; 9: 223-7.
[63] Franklin ME, Chenier TC, Brauninger L, Cook H, Harris S. Effect of positive heel inclination on posture. *JOSPT.* 1995; 21: 94-9.
[64] Nasser JP, Mello SIL, Avila AOV. Análise do impulso em calçados femininos em diferentes alturas de salto. In: Anais do VIII Congresso Brasileiro de Biomecânica. 1997;491-4.
[65] Kulthanan T, Techakampuch S, Bed ND. A study of footprints in athletes and non-athletic people. *J Med Assoc Thai.* 2004; 87:788-93.
[66] Snow RE, Willians KR, Holmes Junior GB. The effects of wearing high heeled shoes on pedal pressure in women. *Foot Ankle.* 1993;13:85-92.
[67] Manfio EF, Vilarde Jr NP, Abrunhosa VM, Souza LV, Fernandez BM, Pereira RM. Alterações na marcha descalça e com sapato de salto alto. *Anais do X Congresso Brasileiro de Biomecânica.* 2003; 1: 87-9.
[68] Aguiar JAS, Freitas TM. Biomecânica da marcha e da postura com calçado de salto alto. *Fisioter Bras.* 2004;5:183-7.
[69] Iunes DH, Monte-Raso VV, Santos CBA, Castro FA, Salgado HS. A influência postural do salto alto em mulheres adultas: análise por biofotogrametria computadorizada. *Rev Bras Fisiot.* 2008; 12: 441-6.
[70] Queiroz AAB, Navarro RD, Kubota MS. Correção da deformidade em valgo do joelho através da osteotomia cuneiforme de subtração supracondiliana do fêmur e utilização simultânea de enxerto autólogo do ilíaco. *Rev Bras Ortop.* 1993; 28:258-62.
[71] Doucette SA, Goble M. The effect of exercise on patellar tracking in lateral patellar compression syndrome. *Am J Sports Med.* 1992;20:434-40.
[72] Loudon JK, Wiesner D, Goist-Foley HL, Asjes C, Loudon K,L. Intrarater reliability of functional performance tests for subjects with patellofemoral pain syndrome. *J Athl Train.* 2002; 37: 256-61.
[73] Aminaka N, Gribble PA. A systematic review of the effects of therapeutic taping on patellofemoral pain syndrome. *J Athl Train.* 2005;40:341-51.

[74] Sacco ICN, Melo MCS, Rojas GB, Naki IK, Burgi K, Silveira LTY, Guedes VA, Kanayama EH, Vasconcelos AA, Penteado DC, Takahasi HY, Konno G. Análise biomecânica e cinesiológica de posturas mediante fotografia digital: estudo de casos. *Rev Bras Ciênc e Mov.* 2003;11:25-33.

[75] Gallagher SP, Kryzanowska R. O método de Pilates de Condicionamento Físico. São Paulo (SP): The Pilates Studio® do Brasil; 2000.

[76] Gallagher SP, Kryzanowska R. The Pilates® method of body conditioning. Philadelphia: Bain Bridge Books; 1999.

[77] Netto CM, Colodete RO. Estadiamento da força desenvolvida pelas diferentes molas do pilates em diferentes distâncias de tensão. *Revista Perspectivas Online.* 2008; 2: 80-91.

[78] Kolyniak IEG, Cavalcanti SMB, Aoki MS. Avaliação isocinética da musculatura envolvida na flexão e extensão do tronco: efeito do método Pilates®. *Rev Bras Med Esporte.* 2004;10:487-90.

[79] Kloubec JA. Pilates for improvement of muscle endurance, flexibility, balance, and posture. *J Strength Conditioning.* 2010; 24:661-7.

[80] Emery K, Serres SJ, McMillan A, Côté AJN. The effects of a Pilates training program on arm–trunk posture and movement. Clinical Biomechanics 2010; 25: 124-30.

[81] Muscolino JE, Cipriani S. Pilates and the powerhouse. *J Bodyw Mov Ther.* 2004; 8:122-30.

Index

C

D

E

M

N

O

P

Q

R

S

T

U

V

W

Y

Z